Supercharging Admissions

with Digital Marketing

for Hospices

Supercharging Admissions
with Digital Marketing
for Hospices

Welton Hong

Founder of Hospice Haven Marketing®
HospiceHavenMarketing.com

Printed in the United States of America

ISBN: 978-1-7378591-1-6

Thanks to my parents Shunhua and Sumei for their guidance, my wife Ihsuan for just about everything, and to my siblings William and Jennifer who make it all worthwhile.

TABLE OF CONTENTS

FOREWORD

Thank you for taking the time to check out ***Supercharging Admissions with Digital Marketing for Hospices***.

Over the last few years, few things have been as gratifying to me as the opportunity to educate small businesses on all the new marketing techniques available in the digital age.

My online marketing agency has helped over 600 small businesses generate more revenue, capture more premium leads, attract better referrals, and set themselves apart from local competitors.

To avoid any confusion, I should note that my principal business is Ring Ring Marketing (RRM), and most of RRM's clients are funeral homes.

With that said, this book is specifically about how we help hospices grow and succeed. You'll most often see the name **Hospice Haven Marketing** (or "**HHM**") throughout this book, and that's no mistake: Hospice Haven is a rapidly growing division of RRM.

HHM exists to ensure we have comprehensive resources dedicated to helping hospices across the country generate more admissions, strengthen referral networks, and better brand themselves in their area.

And as I'll often note in this book, digital marketing has never been more important to businesses of every stripe—and that certainly includes service providers such as hospices.

Even before the pandemic, more people every year—and of every demographic—were looking to the internet to research their options.

That trend was rapidly accelerated by the pandemic. When local lockdowns kept people at home for long stretches, getting information online went from being an option to an absolute necessity.

This was especially true for older people who previously might have preferred offline options to online ones. People in their sixties, seventies, and even eighties—that is, the ones who weren't previously online often—learned how easy and convenient it was to handle many of their affairs on the internet.

Which brings us to now, when virtually everyone from a teenager to an octogenarian grandmother is comfortable ordering products by just talking to an Amazon Alexa Show or pulling up a menu with a QR Code.

I'll provide other examples later, but here's the big point: Your hospice needs to embrace the power of the internet to supercharge admissions and expand your referral network. Today's technology provides a variety of tools to accomplish those goals, and I'll cover many of them in this book.

So please read on, and thanks again for your interest!

~ Welton Hong

INTRODUCTION:

Marketing Is Now (Virtually) ALL Digital

I've written books about digital marketing for several verticals. And in each of the last few — the last being about four years ago — I kicked off the book by explaining that the "Yellow Pages are dead."

Well, they're still dead. That certainly hasn't changed. Unlike vinyl records, they're not coming back to life. And I do want to touch on that a little bit here.

But as I write this in fall 2023, I probably don't need to dig into it as much. I'm going to presume you already know there's no real benefit to advertising your hospice in the Yellow Pages.

The point I was making in those previous books is that people don't look to print directories to choose service providers anymore. And four years later, that sea change is greater than ever.

When it comes to marketing any small business in today's digital age, your primary outlet is the *internet*. In a very real sense, virtually *all marketing* today is *digital marketing*.

Now, I'll grant that even though I run a very successful digital marketing agency, I'm not entirely opposed to offline marketing. I just see zero benefit to a directory such as the Yellow Pages.

Let's get real: A lot of areas don't even receive deliveries of the print Yellow Pages anymore. For people who do receive the print directory, most throw it right into the recycle bin. And they should!

Over the past 15 years, more people have been looking up business services online than in print directories. The trend rapidly accelerated over that time. But the real gamechanger occurred in 2020 with the *global pandemic* and its resulting lockdowns.

It was one thing when younger people — and by that, I mean everyone from preteens to people on the cusp of middle age — took to the internet to find service providers. It was natural for them. Most had grown up with high-speed internet. Their iPhones and Android phones had become constant companions.

Still, some holdouts remained. You know the types. The ones who still preferred reading a print newspaper. The ones who said they'd never give up their landline, that they only had a mobile phone (likely from Consumer Cellular) for "emergencies," or when they were traveling.

The vast majority of these people, of course, were older. They were in their sixties, seventies, or eighties. Many were retired. They'd rather talk on their phone than text. They might have learned to video chat on FaceTime, Skype, or Zoom to keep up with distant family members, but that was about it.

And then… boom. *The pandemic.*

I don't need to expound on that — we all lived through it. But let's take a quick look at the ramifications.

Along with the pandemic came the lockdowns. And due to the lockdowns, suddenly the internet wasn't simply optional for these holdout seniors — it was indispensable.

When you couldn't leave the house — or didn't want to take a chance on getting infected, especially when before treatments and vaccines became available — you had to order products and services *online*.

If you couldn't have groceries delivered to your area, you could order curbside pickup. Which, of course, you needed to do by ordering online and alerting staff with your smartphone when you arrived.

If you did decide to eat out at a restaurant, you had to learn how to use your phone to read a *QR Code* to pull up the menu. (QR Codes appeared to be a dying technology before the pandemic—now they're used in all sorts of marketing.)

I could provide lots of other examples, but here's the short version: The pandemic expanded the widespread use of all digital technologies to an older generation who might have been somewhat technologically literate, but now is fully so.

Sure, there are exceptions. You might have a grandfather or great aunt who still has little use for technology and doesn't plan to change. We likely all know someone like that. But it's atypical.

The reality is that we are all now fully invested in the internet to provide not only services and products but also the research that drives consumers to those services and products.

What does this mean for you?

If your hospice is like most, you likely depend on referrals for many of your placements. And hopefully you already have a strong referral network.

With that said, do you realize how much *stronger* your referral network could be with an aggressive digital marketing campaign? Through many of the tools and techniques explained in this book, you can put your hospice foremost in the minds of physicians and caregivers in your area, helping you generate not only a greater *quantity* of referrals, but ones of greater *quality*.

Additionally, improving your search engine optimization (SEO) and search engine marketing (SEM) will showcase your hospice on search engine results pages (SERPs) on Google and other search engines, showcasing your hospice for referral sources.

(This branding also puts your hospice directly in front of families and patients directly searching for local hospice providers. While I realize direct searching by families/patients remains rare, currently representing a small fraction of hospice lead generation, there are indicators that will change over time [as I'll discuss further later].)

Note that these types of campaigns typically have an excellent return on investment (ROI), and they can improve the likelihood of attracting private pay clientele in addition to standard Medicare admissions.

Further, when you utilize digital marketing tools to improve your *online reviews* and *reputation*, your hospice becomes much more competitive in generating admissions when compared to your local competitors.

Today, virtually everyone checks out Google reviews before considering any type of business, which makes Google reviews very important. CAHPS scores certainly are highly relevant, but at the time of this writing, one must seek them out through Hospice Compare. By comparison, Google makes it virtually impossible to miss its star ratings — they're plainly displayed on any Google search.

That's just the tip of the iceberg. There are huge benefits to be had with social media marketing, retargeting campaigns, hyper-geotargeting, and more. There's never been more potential in the online marketing landscape than there is *now*.

As I mentioned, before we got into the hospice field, most of our clients were funeral homes. And I can tell you, some of those funeral directors originally were skeptical about the value of digital marketing. They had been doing things the same way for decades, and they didn't understand why they needed to evolve in the digital age.

But they figured it out *fast*. They saw that their competitors were generating lots of Google reviews and improving their websites to better convert visitors. They saw how Facebook advertising was changing the game. They realized how remarketing campaigns gave them a "second chance" to convert a visitor who had visited their website.

Next thing you know, they were all in. And that's what I hope for you. I want your hospice to increase its average daily census and be financially successful.

Of course, I realize that's not the only consideration. You wouldn't be in this field if you didn't want to help people. You want to run a successful business while also providing the very best, most thoughtful care to patients.

These elements go hand in hand. The more successful your hospice is, the more patients you can help. More people can benefit from the assistance of you and your caregiving team.

And to continue doing that, *you cannot remain static*. I recently spoke with an authority on hospice businesses who simply said, *"If you are not growing, you are prob-*

ably dying." His point was that the hospice market is becoming highly competitive, including consolidation with larger firms who often have more money and more experience in promoting their firms. To compete, you must *actively promote your hospice.*

When we began helping funeral homes, we knew it was imperative to provide marketing campaigns that were thoughtful and respectful. We understood the sensitivities involved with promoting businesses that served grieving families. And I'm pleased to say we've done an excellent job with that.

Similarly, I'm aware that promoting a hospice is not the same as promoting most business types. (For one thing, physician referrals aren't just part of your lead generation; they're almost all of it — at least for now.) My team understands this. I've built a large staff of marketing professionals who are all trained on how to get premium ROI for clients while delivering messaging that's nuanced, respectful, and caring.

You wouldn't want to trust your branding to a generic marketing agency. Even if they could understand the nuance of promoting a hospice firm, frankly, most so-called agencies out there aren't really agencies!

If you'll forgive me tooting my own horn for a moment, here's what I mean:

Hospice Haven Marketing is an actual full-suite marketing firm. As of this writing, we employ over 125 full-time marketing professionals. We have over 600 satisfied clients who rely on us to promote their services. Every client has their own dedicated coach, who helps them refine their strategy every month.

That's not something you'll find anywhere else. Most firms calling themselves digital marketing agencies are actually just a handful of people.

They set up a few automated processes and no nothing else. They don't create content for you. They don't monitor social media and reputation for you. They don't help you make your website stronger. They don't create high-ROI Facebook and Google ads to generate great leads. *They take your money and effectively disappear.*

So if you've been burned by a purported marketing agency before, I get it. There are far too many bad actors in this space.

That's why we work so hard to earn the trust of clients and maintain that trust in everything we do. It's also why we never lock clients into long-term contracts, and why we provide a no-questions-asked *money-back guarantee* to every client for the first two months.

But that's enough about us. Even if you never consider my firm for your marketing needs, I want you to have the information contained in these pages. Some of these things you can do on your own, and I encourage you to do so! And if you eventually decide you could use something more robust, we'll be ready to help.

So please enjoy the tips and strategies that follow, and don't be shy about contacting us when the time is right. Thank you for your interest!

Section 1:

Lead Generation

CHAPTER 1
Search Engine Marketing (SEM)

If your hospice has a website (and I certainly hope it does), you should have at least some idea of what **search engine marketing** (SEM) entails.

Regardless, let's go over the basics of SEM. In a nutshell, it's an umbrella term for a variety of tools that provide your site the *best possible visibility* online given certain factors.

In other words, you're obviously not competing for visibility against every other site in the world. No matter what you do, you're not going to be more visible than Amazon for retail goods or more visible than iTunes more music sales.

In SEM, you want your site to be more visible than the sites of local competitors in your field. As a local business, that means optimizing your site (and your marketing efforts) to rank higher in search engine results pages (SERPs) than other hospices in your market and in outlying areas you serve.

These days, online search is extremely specialized. There are strategies and tools that let you put your best foot forward in a particular area and market.

Some of these tools essentially require no advertising dollars; by optimizing your hospice website with particular keywords and types of content, the site will organically rank well in your market and area.

Other tools require dipping into your marketing budget, such as letting you bid on particular keywords that people who are looking for hospice services — and potential referrers, of course — will be using when they search.

For your hospice, this likely would include various forms of keyword phrases including targeted keywords that should include your city and state, along with terms such as *hospice*, *palliative care*, *end-of-life care*, etc.

A successful internet marketing plan typically combines both approaches, one that balances the initiatives of organic SEO and paid advertising.

In this chapter, I'll cover the basics of **pay-per-click advertising** (PPC), **local search optimization**, and **search engine optimization** (SEO).

Pay-Per-Click Advertising (PPC)

Pay-per-click (PPC) is an umbrella term for several types of advertising, such as sponsored listings, paid search, partner ads, or sponsored links — with one thing in common:

In each type, you pay for the ad only when a user clicks it and is forwarded to your website (or whatever destination page you set up).

Most often, you'll find PPC ads atop pages when you do online searches — you'll recall those pages are called SERPs, or search engine results pages.

Google remains the dominating industry leader in search, with over *90 percent* of the organic search market share as of this writing. (Yes, it's *that* dominant.)

Its PPC function is an aspect of **Google Ads**, the venture that provides Google most of its revenue.

(A quick note: Google's parent company is technically "Alphabet," which is essentially just a renaming of its corporate entity.)

One of the coolest things about PPC ads is that they don't appear until someone completes a search.

Because they're dynamic, the ads that appear following the search will be relevant to whatever the user is searching for. These ads are, by definition, ***targeting an interested party***.

In other words, if someone (either a referrer or family caregiver) is already searching for topics related to hospice, that person will see these ads—they won't display to people with no relevant interest, helping you avoid worthless clicks on your ad.

The obvious benefit is that the ad doesn't cost you a penny simply for appearing: *You only pay when it gets clicked*, sending the user to your page and providing an opportunity to convert the visitor into clientele.

Creating a PPC account can differ slightly depending on the platform (an internet marketing professional can explain the process and handle it for you). But here's the general game plan:

First, you'll create a PPC account with the particular platform, a simple enough process. Next, you'll create campaigns and ad groups.

The campaign part is where you determine options such as the geography of the market you wish to reach. Because your hospice relies on local residents, for example, you'll obviously want to target local people.

That's how localized search and advertising is these days: You can set it up so *only* people in your city, county, or state (depending on your preferences) see your ads and thus click on them.

The ad groups comprise your advertisements and keyword lists. An internet marketing professional can help you determine whether there are any other specific terms people will be searching for when they want to find a business that provides your services.

Remember that despite the name, *a keyword isn't necessarily just a single word*; it's very often a phrase. When a keyword phrase comprises more than one word (typically three or more), it's commonly called a **long-tail keyword**. You'll incorporate these keywords in the ads you run in your campaign.

The search engine for that campaign (typically Google) will determine how well the advertiser's campaigns and keywords match the user's particular search, and it will display the ads if there's a close match.

A few other factors determine whether your ad will show up. Every search engine has an algorithm that considers these factors and employs them with each search. These factors include:

- How closely do the keywords in the list match the actual search?
- Does the advertisement itself include a mention of what the user is looking for?
- Does the ad forward the user to a landing page selling the service the user wants?
- How much are you willing to pay for a user searching for these terms to see your ad (and hopefully click through to it)?

The final point — the actual amount of your bid — determines not only how much you'll pay, but also *how prominently* your ad is displayed on the page.

The specifics of each search engine's algorithm are not made public for a sound reason: It keeps less legitimate advertisers from gaming the system, crafting ads that exist only to get clicks without actually providing the user anything useful.

Google, in particular, has tweaked its algorithms substantially over the years to reward websites and advertisers who are legitimately providing good information and selling useful services and products. This is happening not only in PPC but in search engine rankings; we'll talk about this some more later.

The thing to remember is this: If you provide good content on your hospice's website/landing page and craft good PPC ads to market your hospice, you'll be rewarded.

However: If your sites and ads are intended to mislead people into clicking through just to increase site traffic, you're taking a chance on being *penalized* by search providers, particularly Google.

This is one of the most common **black-hat** tactics. Anything you see referred to as "black hat" is a nefarious attempt to game the system. I can assure you, it's never worth it to employ black-hat tactics. No truly reputable marketing company ever does.

If Google or other search engines catch you trying to game the system, your hospice's entire digital presence could be massively penalized. *It's not worth the risk.*

Instead, only work with companies who employ **white-hat** techniques. As you've surely inferred, that means doing things the right way.

Whatever short-term benefits a business might receive from black-hat tactics will be far outweighed by the massive damage it can do when Google sniffs it out. And Google *will* sniff it out: It's like the IRS — just much faster and smarter.

Whether PPC is a valuable investment for your hospice depends on several considerations. A big one is whether your website/landing page typically shows up near the top of the results when people are searching for services you provide in your area.

As anyone can tell you, many people don't bother searching past the first page of SERPs. (Think about it: How often do you continue to Page 2 or 3 after doing a Google search?)

Being ranked on that first page is critical. If you're not ranked for whatever reason — perhaps your site isn't sufficiently optimized for search engines, or you're in a highly competitive market — you're virtually invisible to interested parties even if you're near the top of the listings on Page 2.

If you're even deeper in the SERPs, you're essentially *invisible*. Most searchers, if they don't find what they're looking for by Page 2, will just try a new search with some different keywords.

Even better than simply appearing on the first page of Google search results, you definitely want to appear in the **Google 3-Pack** (alternatively known as the Three-Pack, Map pack, or Local pack).

The 3-Pack is a dynamically displayed list of firms displayed prominently on the first page of search results. Again, the contents of the pack are directly related to what the user was searching for. We'll dig into that more later.

A PPC ad lets you essentially jump ahead of the SERP organic listings, putting your hospice's website or landing page right out front. Granted, this comes at a price, and some searchers remain reluctant to click on an ad as opposed to a search result. (Note that this stigma against search ads has diminished in recent years.)

However, PPC remains an *extremely effective tool* for many reasons. It provides a lot of control and is highly customizable. You only pay if the ad is clicked.

PPC also makes it easy to get extremely accurate statistics on whether your ad is effective, unlike most traditional types of advertising.

Best of all, if your hospice needs to *rapidly increase leads immediately*, PPC is the perfect tool for the job. PPC ads are highly visible, and they're especially well displayed on mobile devices.

Smartphone users can call your firm with a single click on a mobile PPC ad. And that's one of the biggest keys to digital marketing success: *Make it easy for a potential referrer or client to contact you.*

Whether you get the call or a competing hospice does often comes down to simply which one put up fewer obstacles for the searcher. *When you make it easy, you get the call.*

The analytics of PPC ads let you know which keywords people used to find your page, so you can tweak your keywords depending on which ones worked well (or poorly).

You'll know how many people departed a page on your site immediately after reaching it (known in the industry as your **bounce rate**) — and how many stuck around.

One of the biggest benefits of PPC (aka paid search) advertising is that it *bridges the time gap* between SEO improvements to your website and when they're recognized by Google.

Improving your organic SEO (the right way) goes a long way toward elevating your search rankings for people searching for hospice services in your area.

However, Google only "crawls" your site (more on that later) occasionally, so it won't know immediately that you should be ranked high on the first page of results.

By comparison, PPC ads post almost immediately, letting you get your name right out front while you're waiting on Google to review your site and increase your organic ranking.

Some other benefits in using PPC advertising to hone in on the right demographics and target the best leads:

Geography: This is a key factor in ensuring that you're targeting the most qualified leads. Analytical information lets you tweak the geography of your ads to concentrate on markets where they're doing well (and lets you pull them from markets where you're not garnering interest).

Similarly, one of the most useful aspects of PPC advertising is that you can target potential clientele in very specific geographic areas.

For example, a PPC ad can be set to only appear *within a set radius* of a firm's physical location. This is a great option for a hospice, obviously, because it's a local service.

In addition to radius, PPC ads can be set to run only in particular ZIP codes or cities. Whatever exact areas you wish to target, your PPC ads can be focused entirely on them.

If you wish to cast a broader net, no problem. Unlike physical media such as the Yellow Pages, it costs *nothing extra* to target multiple cities.

You're not paying five or six times as much to run an ad in five or six cities —*one price fits all*. Thus the cost per lead is far less expensive than it would be with traditional media.

Day parting: As with geographical targeting, PPC ads can be set up to run only during the times of the day or week you want them to run.

Traditional service/product businesses most often stick with running ads during their operating hours (or perhaps extended to an hour or so before opening), but there's certainly a competitive advantage for a hospice to run PPC ads even during nontraditional hours. After all, you should appear available to accepting leads and referrals 24 hours a day. I'm sure you have staff on call 24/7, so be sure to communicate that in your hours!

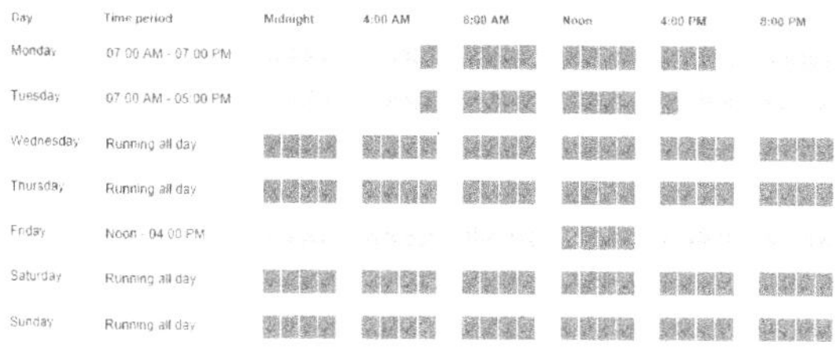

(Quick note: I just did the very simple experiment of searching "hospice near me" to get a list of local providers. Some list their hours in their Google Business Profile as 24/7, while others only show standard hours. I can assure you, the ones that never appear to be "closed" are much better positioned to secure market share.)

Controlling lead flow: The scheduling of PPC ads can be easily modified to increase traffic. Google charges based on a daily budget.

Let's say you typically spend $30 per day, but there's a period where you need to increase your new business flow and have some room in your budget. You can increase it to $50 for more coverage.

On the other hand, let's say your hospice is already very busy at the present time. You could knock down your PPC ad budget to $15 for a while to save some money and ensure you don't get overwhelmed. (If you're busy already, why spend money on ads?)

And if you're literally *overwhelmed* with business at a particular time — a good problem to have, but a challenge nonetheless — you can even *pause your ads* for whatever time frame you determine. This also comes in handy if your hospice, for whatever reason, is closed for a period of time.

In this sense, running PPC ads is like turning the knob on a faucet: You can open it up to flow freely, turn it down to a trickle, or turn it off entirely for a while, all dependent on your needs at the time.

Instant traffic: As I noted before, PPC ads let you be seen on the first page of search results almost immediately. *You're paying for the opportunity to jump to the front of the pack.*

This makes PPC a powerful tool for seeing a very fast return on an online marketing campaign. When used in tandem with organic SEO and local optimization strategies, you will see your traffic soar and watch it continue climbing in both the short and long term.

PPC ads are never static: Anything can be tweaked, adjusted, or swapped out at any time. With print media, you're stuck with the same print ad for a long time. In the case of the Yellow Pages, you're stuck with it for an *entire year*.

If there are changes in your business — you add a location, you change the geographical area you're targeting, you revise your services in some way — the Yellow Pages ad will still have all the old information until it's time for a new edition of the print directory.

(Note: While I absolutely believe some offline marketing can be beneficial, you can tell I see virtually no reason to pay for a print Yellow Pages ad in the 2020s.)

PPC ads let you update content on the fly. If you want to make a change at any time, you can have the new ad appearing "live" to searchers in only a day or two.

Trying lots of variations: With a typical print ad, you come up with terms and phrasing you think will work best, cross your fingers, and hope it does the job.

PPC advertising lets you try out many different variations of ads catered to the keywords you want. You literally could have hundreds or even thousands of different variations employed at the same time.

(To keep things easy to manage, you'll likely only want to employ a handful of keywords, but you get the idea.)

This also allows for **split testing** of the different choices, letting you know quickly which ads have the most impact and which are less successful.

Additionally, when you have several ads running simultaneously with differently worded long-tail keywords, the ads people see will typically match up with the terms they're specifically searching for.

This is another way in which PPC advertising beats out print, because in something such as the Yellow Pages, you have to determine which sections you want to be listed in and place ads in *all* of them.

Excellent branding tool: If improving brand awareness is a critical concern for your hospice, PPC is an effective tool. If people click on your ad, that's great: They're coming to your website to consider your services. If they don't, you're *still getting exposure* to lots of people (clients or potential referrers) searching for what you provide, improving your brand awareness.

Note: Now, I know that terms such as "brand awareness" can sound a bit too "business-oriented" to many hospice professionals.

You're not a *brand*, you're a hospice, right?

Well…not entirely.

You actually are a brand.

Of course, you provide essential services for people in need of care. Of course, that makes you very different from other businesses. You don't sell pizzas. You don't fix transmissions. You help patients and families in need.

I get that—as does my entire staff, all 120-plus of them. Everyone at Hospice Haven Marketing understands the distinction between a critical service provider and an everyday business.

Regardless, you are a business, and branding yourself is critical to generating more admissions, improving your referral network, and helping more people.

I realize it's important that you maintain your identity as a benefit to your community, and I'd never suggest otherwise. You *are* a benefit to your community. But you're *also a business*.

So yes, it's perfectly fine to think of yourself as a brand. Just think of yourself as a particularly *beneficial* brand.

After all, if you're that good at helping patients and families in your area, aren't you helping *them* out by ensuring they can easily find you and work with you? Are they well served by working with a competing hospice that can't serve them as well as you do?

You're a brand now. Accept it and incorporate it into your marketing strategy. It's the best thing for you *and* your community.

You don't necessarily have to pay for the top spot in PPC ads: Consider the NFL draft. Teams regularly trade down from their initial spots in the draft because the value isn't there for the player they need in a particular position.

They often can obtain that player even when they pick a little further down in the draft — and ultimately *pay less money* for the player, because the spot where a player is picked goes a long way toward determining his salary.

(Okay, football fans, fine: In the draft, teams also can get more draft picks for trading away a high pick, but that part's not relevant here.)

The same is true of PPC advertising. For some organizations, and this includes hospices, bidding for the top spot can cost much more than the value it delivers. The top spot for a particular keyword typically is much more expensive than, let's say, the fourth spot.

Having that top spot makes sense for a business for which people will likely not take the time to perform any additional research — for example, a locksmith or a bail bonds provider. For those types of businesses, people often click on the very first ad that pops up when it's an emergency or time is of the essence.

In that sense, getting a higher position might be worth the investment for your hospice. If it's critical for you to be the *very first option* families see when dealing with an unexpected death, it's worth paying the extra freight for that top spot.

However, when it comes to care services that tend to be researched more extensively (because a decision doesn't necessarily need to be made immediately), there's less benefit to paying for the top spot. In that case, an interested party likely will be checking out several providers that pop up for those keywords.

In that situation, ranking third or fourth is just as good as ranking first, and it's substantially less expensive.

PPC also can include display ads: While the focus of this discussion has been PPC ads that appear on search engine results pages, you can also publish display ads on the pages of Google's display ad partners. These are the types of ads you commonly see on larger websites.

In general, advertising for SERPs is intended to attract people who are looking for services immediately. Display ads on larger websites are more intended for branding and to attract people who might be interested in your services down the line — so that's best for growing physician referrals or attracting people who might expect to need hospice care sometime in the future.

The process for publishing display ads on Google is known as the **Google Display Network**. Google has "flexible reach" targeting, which ensures greater control over targeting and the ability to try out different targeting combinations more easily.

Remarketing: This is another great benefit of PPC advertising.

When you employ Google Ads, you also can leverage remarketing, which means that your ads will *continue to make a case for your hospice* with people who have shown an interest in your services by visiting your site previously.

Google can track who has visited your site and continue to display your ads for these leads when they visit other sites. This lets you reengage with leads who did not purchase immediately.

You can even customize the messaging in this ad to make a special offer that could compel them to return to your site and work with you.

Remarketing has been hugely successful in recent years across a wide variety of services. This marketing strategy can attract people who are planning ahead for hospice services.

An important consideration in remarketing is that you want to set the ads to run only for a particular time frame or to appear only a certain number of times. That way, remarketing targets won't feel like they're being "stalked" by your ads all over the internet. No one wants that, and it would be bad for your hospice's reputation.

Thus, remarketing must be employed surgically, and a good marketing firm can help you with figuring out the proper time frames and/or number of appearances.

Summary of PPC advantages

Virtually instant traffic to your advertisements

- You can have ads on the front pages of Google (and/or Bing and Yahoo, if desired) in just few hours.

A world of options print Yellow Pages can't provide

- Use text-based ads, streaming video ads, banner ads, and more.
- Advertise to a targeted audience in minutes.
- Be rewarded for creating good ads with lower ad costs.

Laser targeting

- Advertise to people who are looking *right now* for the services you offer.
- Quickly test to determine which ads bring you qualified traffic.
- Easily monitor which keywords attract visitors who actually convert into clientele and which do not, letting you easily alter your ad to include the highest-converting keywords.

Local/regional targeting

- Target users by location, allowing you to specifically reach local clientele.

- You can choose where and when you want your ads to be visible.

Cost-effectiveness

- Pay only when an interested searcher clicks your ad.
- Benefit from free branding even when people don't click your ad.
- Determine your costs based on your own objectives: Specify exactly how much you want to spend, and even set a maximum daily and/or monthly budget.

Measurable results

- You get real-time ROI data through conversion tracking data.
- Every aspect of the process is trackable.
- Google collects and analyzes all key data for you and will auto-send you reports.

Summary of PPC disadvantages

- Only through paying the advertising cost will you be able to determine the ROI of various ads.
- Each new visitor costs money: While you could get some return visitors who bookmark your site, in general when you stop paying, the visitors stop coming.
- Having your ads link to a high-converting site is a must: A good PPC campaign can drive lots of targeted traffic to your hospice website, but if

that site doesn't turn visitors into clientele, it's wasted money.

A note on the landing page

With all that said, a PPC ad campaign is *only as good as the landing page it's selling.*

Let's say you're looking for a home and you drive past a gorgeous billboard advertising an amazing condominium complex just a mile away.

It's advertising resort living at a fraction of the price, tons of amenities, and a great neighborhood. It sounds like paradise.

So you head over to the address, full of excitement and glee, and when you arrive, you discover... it's a dump.

It looks like it hasn't been renovated in 100 years, it has a rusty, broken gate out front, and the parking lot is covered in trash that's clearly been piling up for weeks.

Are you going to bother driving onto the property and checking it out? Or are you going to get out of there as fast as you can before you get mugged?

That's what happens when a landing page doesn't deliver what the PPC ad promised.

If the landing page is poorly designed, or if the website isn't selling what the typical visitor would expect, the visitor is going to "click out" in a matter of seconds. And you just paid for a visit that did you no good.

I'll get deeper into landing page and website optimization in a bit, but it's critical to remember that there's no point to purchasing PPC ads if you're sending visitors somewhere they don't want to be.

Anything from a lack of relevant content to slow page loads can prompt a visitor to set sail in a matter of seconds.

The bottom line is this: Your PPC ads are only as effective as your website is. Your SEO efforts are only as effective as your website is. Your offline advertising… well, you get the idea.

Just as there would be no point to spending thousands on marketing a pizza parlor with terrible food, health department citations, and a lazy waitstaff, there's no point to driving families to a hospice website that has poor navigation, no obvious call to action, and little information on the firm and its services.

The pizza parlor needs to focus *first* on cleaning up the kitchen, improving its food, and hiring better servers. The hospice site must focus *first* on better serving its visitors and converting them into clientele.

Only then will PPC ads and other marketing methods generate more leads, admissions, and referrals.

A site can attract tons of traffic, but if those visitors don't turn into clientele, the traffic is pointless. Once you've transformed your site into a conversion-generating machine, PPC ads will work wonders.

Local Search Optimization

In the most basic sense, when we're talking about local search (you'll sometimes see this capitalized as Local Search, but I'll stay old-school here), we're simply referring to online searches that have a geographical component.

If you live in Tallahassee and need to have a replacement odometer installed (by a trained mechanic) for your 2017 Ford Fusion, you typically won't simply search for "2017 Ford Fusion odometer" or "Ford Fusion mechanic" or "2017 Ford Fusion repair."

You'll usually add something that *includes your actual location*. Thus, you'll add "Tallahassee" to your search terms.

With that said… on most devices, the search engine will *already know where you are* and automatically deliver local results, even if you don't add a location. This is usually true as well if you include "near me" in the search.

Recently, search has become increasingly local. Again, this owes a lot to the massive expansion of affordable mobile technology. Smartphones and tablets (and more recently, even *smartwatches*) let people navigate their neighborhoods easily.

As of fall 2023, Google noted that ***46 percent*** of searches have local intent. When you limit "local intent" to mobile (specifically smartphone) searches, that number eclipses *80 percent.*

It's actually rather ironic: In its early days, the internet let us quickly and easily interact with people and places all around the world. These days, we're more likely to use it to find places and services *right in our hometowns.*

Of course, that's why local search optimization is so critical for hospices in 2023 and beyond. As I noted in the introduction, this is how many potential clients will find your hospice these days.

Optimizing your hospice for local search ensures not only that you're putting your best foot forward for potential clients — it ensures that you *show up at all*. You need to do this to show up on Google Maps, the most widely used graphical search provider in the world.

Practically every day, I find myself needing to find something near my current location — an ATM, a gas station, a restaurant, a grocery store.

My phone automatically knows where I am, thanks to GPS. Businesses that are optimized for local search will pop up as soon as I search, and the listings show me how far away they are, directions, etc.

But that's not all: When a business has included its business hours on its **Google Business Profile** listing, I automatically know whether it's open or closed — or how long I have to get there until it *does* close. (As noted earlier, I strongly recommend setting this at "24/7.")

I can also see ratings and reviews, helping me know whether I can trust that particular mechanic, hair stylist, or retailer.

This is one of the key processes people now use to find local services. If you're not taking advantage of it — if your hospice hasn't fully filled out its Google Business Manager, or if the information there is out of date or incomplete — you're placing yourself at a massive disadvantage compared to your competitors.

Getting to know Google Business Profile

It's useful to understand a little history here:

Google has been focused on local businesses for close to two decades. In fact, what we now know as the Google Business Profile only got that name in late 2021. It's been through several previous iterations. Here they all are, with the year Google named them:

2005: Google Local Business Center
2009: Google Places
2011: Google+ Local
2014: Google My Business
2021: Google Business Profiles

There's a lot more to it than that, but let's focus on what matters. Google Business Profile is now firmly entrenched as "*the* Google tool for small business."

As you surely know, people do review hospices on Google, Yelp, and other sites. Having the strongest reputation in your area immediately makes you a frontrunner for these services.

Recent surveys show that over 77 percent of consumers consider online reviews as trustworthy as personal recommendations, which is why ensuring you have positive (and preferably exceptionally positive) reviews on your hospice's GBP listing is so important.

When dealing with businesses in other industries, this is where I typically would note that it's a good idea to ask satisfied clientele to review your firm positively.

And yes, this is where we get to the potentially sensitive spot.

As I mentioned way back in the foreword, my agency worked with (and continues to work with) hundreds of funeral homes before we started helping hospices.

A lot of funeral directors absolutely *hated* the idea of asking for reviews. (Some still do, to their detriment.) They worried they would seem insensitive, that they would appear to care more about their bottom line than about caring for bereaved families in their community.

Thankfully, we were able to convince our clients that asking for reviews is just fine—and believe me, all our deathcare clients are much better for it.

I certainly appreciate that asking for a review from a family whose loved one was cared for by your hospice requires thoughtfulness and nuance. It's not like asking for a review of a restaurant or an electrician. You need to show caring and grace.

Here's the thing: **You already do that!** You sp*ecialize* in that! You know how to help patients and families through incredibly challenging times. You know how to be thoughtful while also running a business. So you're already perfectly suited to graciously ask for reviews.

We conveyed a similar message to our client funeral directors, and sure, some were skeptical at first. However, the ones who wasted no time getting on board saw immense returns from their initiative.

Your online reputation is **massively important**. At the risk of stating the obvious, no one wants to entrust the care of their loved one with a hospice that has poor reviews (or very few reviews) on Google. Even if they got a very strong recommendation from a physician, nurse, or other care provider, some families are going to check out the hospice on Google. (This will continue to grow over time.) And if the reputation they find isn't strong, many will consider other options.

At Hospice Haven Marketing, we also utilize a proprietary email program that solicits reviews from hospice-served families in a very thoughtful, understanding manner (after a proper waiting period as appropriate).

It's another good option for hospice professionals who worry about asking for reviews in person.

Soliciting reviews is just like anything else in your field: It needs to be handled with the same sort of sensitivity and dignity you're used to providing and ensuring.

A quick sidebar: Of course, this is separate from the standard CAHPS Hospice survey, and I appreciate that you wouldn't want to ask too much of a grieving family. Spacing out the timing of a Google review request from the CAHPS survey is obviously optimal.)

Your GBP listing also effectively acts as a blog, thanks to the **Google Posts** feature. You can provide updates and reach out to families and referrers, providing useful information and showing potential clientele you're highly engaged and communicative.

Publishing short updates on Google Posts also helps with your hospice's website SEO (because Google sees you more as an influencer in your industry), so you're getting two benefits every time you post!

Don’t forget Bing Places for Business

Of course, Bing has its own local option, and while it makes sense to make Google Business Profile your top priority, that doesn’t mean you should entirely ignore Bing.

While Bing comprises a comparatively tiny portion of the search market — speaking of great branding, there’s a reason you never hear someone say “just Bing it” instead of “just Google it” — Bing still plays a role.

And that’s simply because Bing is Microsoft’s search engine, and Microsoft remains an enormous player in computing, thanks to its popular Windows software.

That’s why when you buy a new laptop or desktop computer with the Windows operating system, the default search engine (on Microsoft’s default web browser, Microsoft Edge) is Bing.

Granted, most people with even a modest understanding of technology know they can change their computer’s default web browser in just a couple of minutes. (It’s incredibly easy.) So if they prefer a browser such as Google Chrome to Microsoft Edge, as many do, they just make the change.

(Note: As of fall 2023, Chrome had 68 percent of the web browser market share, compared to slightly less than 6 percent for Microsoft Edge. Safari, the default browser for Apple operating systems, had 15 percent.)

Additionally, virtually all smartphone users in the United States either use Android devices (Android is a Google operating system) or iPhones (Apple).

As of this writing (fall 2023), about 43 percent of Americans used Android phones, while 57 percent used iPhones.

Regardless, there's no reason not to ensure your hospice is properly listed on Bing Places for Business.

Your hospice has a physical address, so it likely already has a Bing Places for Business listing; you simply have to claim it.

Much like Google Business Profile, you'll want to ensure your profile is completed with all the pertinent information for your business.

Note that if you haven't filled out your Bing Places for Business listing previously, you might have an even more convenient option. Bing actually lets you *import* the information from your GBP account, if you wish. (This won't affect your GBP listing.) It's right there in the setup process.

Just make sure that you've checked all the information in your Bing listing after the import. Don't just assume it's all correct.

Unlike GBP, Bing Places for Business doesn't have its own native reviews. Instead, it pulls reviews from other platforms, such as Yelp and Facebook.

There are some other differences between these two platforms, and it's useful to note that Yahoo has a local platform of its own, but most of the points noted earlier about Google Business Profile relate to all the services.

The most important takeaway is that completing all your hospice's information on these platforms — and ensuring it's accurate and up to date — is necessary for putting your best foot forward. These services are built to accommodate mobile users, a segment that has grown massively just in the past couple of years.

Citations

Another very important aspect of local search optimization is the use of **citations**. These are mentions of your hospice that appear on the internet regardless of whether there's an actual link to your website.

To superpower your local SEO, you must generate a substantial number of local citations.

A local citation could occur when your hospice is noted in general text in an online listing of firms in your local market or your region. It could be a mention in a news story, white paper, or trend piece.

Although these references to your business don't include a direct link, they are key components in the ranking algorithms of Google and other top search engines.

The more you're referenced on other sites, especially if they're among the more regularly visited and indexed sites, the more "useful" the search engines think your hospice is. This improves the rankings of your website.

A citation on something such as a chamber of commerce or county business index is especially useful. This is also true of a citation in a major newspaper or other news site.

That's because these citations lend even more legitimacy and credibility to your site — it's awfully hard to "fake" an appearance in a major newspaper, so the search engine knows you're an actual operating firm.

Your internet marketing specialist can help you learn how to improve your citations, but in general, the key is to ensure you're properly listed in all legitimate listing services that are appropriate for your hospice.

Be sure you're included in local blogs, locally focused directories, and directories/blogs that are specific to hospice and palliative care.

If you are involved with something newsworthy that can get your hospice named in a newspaper story or on a local television station's website, that's excellent. Google and other search engines prioritize these types of citations.

Local search advantages

- It's free, which allows for infinite return on investment.
- It lets families rate and review your business, so if you're good at what you do, you'll stand out from the crowd.
- It adds your site to Google Maps, one of the top ways people discover local businesses now.
- It lets you add specific keywords to your business description, giving you a leg up on competitors who have not done this.

Local search disadvantages

- Only a handful of businesses automatically show up as a search result.

- It can take greater than a month to show up in the listings, and the likelihood of appearing high in the results depends on optimization.
- You can only optimize your firm for the five to ten most relevant keywords.

Search Engine Optimization (SEO)

An entire book can be written about search engine optimization (SEO). In fact, many have been, and many more will be. There are many facets to SEO, and the best practices are regularly changing. Frankly, they've changed substantially just in the four years since the previous edition of this book.

Here, I'll touch on the highlights.

You undoubtedly know the purpose of SEO: It's a set of processes intended to help ensure your website or landing page ranks as highly as possible in search results.

Ever since search engines began appearing on the internet, website owners and webmasters have understood the power of ranking as highly as possible on SERPs.

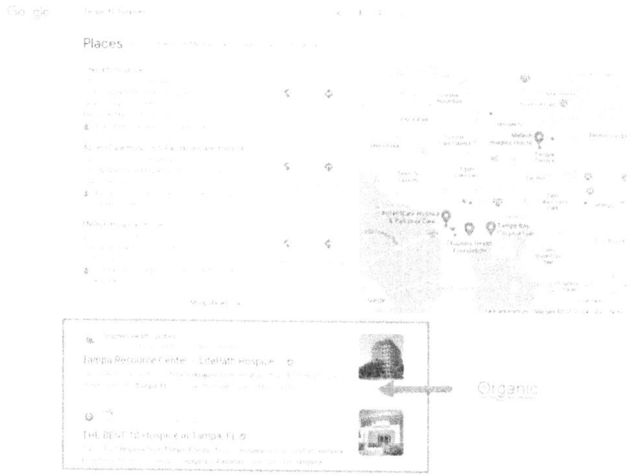

It's not exactly a new concept: Everything else being equal, you're more likely to get referrals and clients if *yours is the first option they see*.

That's why print magazines (yes, a few still exist today) want to be at the *front* of the racks in physical bookstores (if you can still find one of those today — it's not easy).

That's why businesses for many years would select names that put them at the front of white pages and Yellow Pages listings:

Ace Plumbing was upstaged by Abacus Plumbing, which was eclipsed by Aardvark Plumbing, then A Aardvark Plumbing, followed by A AA Aardvark Plumbing… I'm sure you remember how that went. You'll still see a lot of local businesses today using this tricky naming process.

However, today *alphabetization doesn't matter much*. Search engines don't care whether your business name starts with an A or Z. They take no notice of whether you have a really colorful name or a bland one — that's not part of the equation.

Granted, your hospice's name *does* matter, but only in the sense that the SEO will be improved if *terms in the name are highly relevant* to what people will be searching for.

That is, if instead of providing hospice services, you happened to sell custom hubcaps in Charlotte, then naming your business Charlotte Custom Hubcaps (and snagging that as a domain name as well) would be a big help for your SEO.

If you're a hospice in Tucson, obviously, something like "Tucson Hospice" makes things easiest for search engines and potential leads.

Many hospices use "health" in their name, which isn't bad, but it's not nearly as good (for SEO) as the specificity of a name that includes "hospice" and/or "palliative care."

Granted, if your hospice has a well-established name and brand identity in your area, you might not want to consider a name change, which is certainly understandable. However, if your hospice is struggling or you feel your brand would not be too adversely affected, a rebranding could provide great benefits.

Before we dig deeper into how SEO works, let's start off by defining a few terms:

White-hat SEO vs. black-hat SEO: I touched on these terms earlier, but here's a little more context:

As soon as webmasters realized that search engines were using certain elements of websites to determine how highly those sites should rank, some started to think of ways to game the system.

When a site is using tricks that are only intended to improve rankings or traffic without providing an equivalent benefit to actual visitors, that's black-hat SEO.

On the other hand, when a site uses techniques that improve rankings and traffic that are integral to providing a benefit for visitors, that's white-hat SEO.

Sneaky sites (and their sneaky webmasters) benefited from black-hat SEO for many years, but those days are largely over now. Search engines (particularly

Google) tweaked their algorithms substantially to punish sites that used black-hat tactics and reward ones that followed a white-hat approach.

As I noted earlier, it's *absolutely not worth it* to work with a firm that promises big results through black-hat practices, no matter how cheap the price.

The money you save up front by working with these con artists — and that's truly what they are — won't come close to making up for the business you lose when Google banishes your site to page eight of its SERPS.

In extreme cases, Google can even *ban your URL altogether*. And once you're on Google's bad side, you're in serious trouble: Google has a very long memory.

Spiders/crawling: With rare exception, search engines don't employ people to actually visit websites and determine whether those sites are relevant and useful (and thus worthy of being ranked highly). Instead, they create software known as **spiders**. These spiders go to websites and then read and index the information there, a process known as **crawling** the site.

They don't just review the content visitors can see, but also information in the coding of the site. The info they collect is then factored into complicated algorithms that determine how highly a site should be ranked for various searches people do online.

Organic search results: Otherwise known as natural search results, these are the basic unpaid search results — as opposed to paid listings, "sponsored" listings, or ads — people get when they search for keywords.

As you've surely noticed, searching for anything on Google these days will result in a page of SERPs that include both organic and paid results, and sometimes it can be tricky to tell the difference.

Ultimately, your hospice services website has two very distinct objectives.

One objective is to put your best foot forward for actual human visitors. You want visitors, whether they're referrers or potential direct clientele, to find you professional and authoritative and select you.

Your goal is to create *conversions*, which essentially means that your site prompts the visitor to take an action toward becoming clientele (or a good referral source).

Most often, you'll want the visitor to either call you, email you, or fill out an online form for more information. If the visitor takes any of these actions, you could consider that a conversion.

(Of course, that doesn't necessarily mean a visitor ultimately will contract with you for services. However, as long as your website prompted one of the actions noted above, the site properly *converted* the visitor.)

To this end, you want the content on your site to read well, provide useful information, and sell the visitor on the fact that you're a trustworthy and thoughtful provider of hospice services.

Beyond dealing with human visitors, the other objective of your hospice website is to sell the relevance of your site to the search engine spiders.

This is where SEO strategies come into play. By employing these strategies in the site's content, design, and coding, you can persuade these spiders to rank your site highly when people search for keywords that would be naturally associated with hospice services.

As you might expect, one of the greatest concerns Google and other search providers have is that sites try to game the system to make the *spiders* happy without making actual human *visitors* happy.

This is why Google, in particular, has started heavily penalizing sites that use this approach. It wants the sites that rank highest to be the most useful for *actual people* who are actively searching for what you provide.

You might wonder why certain sites would game the system with pages that rank highly but provide nothing good for the visitors. People will just click out immediately when they realize there's nothing there for them, right?

Some site owners/webmasters do this because their business model simply pays them for **hits** — visits to sites, whether or not the visitors bounce out immediately.

These sites use advertising models where it doesn't matter at all whether the site itself is useful, just that they've recorded hits for advertisers to the site.

If a visitor goes on to click on an ad on the site, the site gets paid even more. Ultimately, these sites are not really selling anything or providing any service; they exist only to trick people into clicking through to the site.

These are the types of sites Google has penalized with tweaks to its algorithm. Google literally changes its search algorithm as many as *500 times* in a given year. Doing so ensures that people searching Google will find sites that provide the information and products they want.

This provides a great opportunity for legitimate service providers such as you. By combining high-quality content with legitimate white-hat SEO techniques, you make *both* your human visitors and search engine spiders happy.

When your hospice's written copy is relevant, easy to read, and compelling — not just stuffed with irrelevant keywords to trick the spiders into ranking you more highly — you'll actually be increasing your SEO and receiving higher rankings (and thus increased traffic).

It's a lot harder to game the system now, and that's not a bad thing at all: It means legitimate websites such as yours are rising to the top while the tricksters are relegated to the depths of the SERPs.

At the risk of belaboring this point, I need to be clear: Sites that use black-hat tactics are massively penalized when Google figures out what's happening. (And Google *always* finds out.)

They can plummet so deep in search rankings that no one will ever see them, and when the tricks used are particularly egregious, Google (and Bing, Yahoo, etc.) can even remove the site completely from search rankings.

If an internet marketing "expert" claims he or she can substantially improve your hospice website's rankings or traffic without making substantial improvements to the actual content, design, and relevance of the site, don't walk away — *run*.

Trying to trick Google is a very, *very* bad idea. Instead, find a legitimate internet marketing pro with a great white-hat SEO record.

Great original content is great, but sham experts will rip off content from other sites and use it on yours. Relevant incoming and outgoing links are great, but these scammers will add links from hundreds of irrelevant websites.

They'll hook you into massive link exchange sites that provide nothing useful for your actual human visitors. The brief jump your site might see in rankings is not worth the damage that will occur when the hammer comes down.

Doing it the right way

So how do you do it the right way? It's important to understand what a search engine is looking for when determining how highly your website or landing page should rank in SERPs.

Though this relates to all major search engines, let's just use Google as an example. Every time someone does a search for certain keywords, Google's search engine considers the following:

Site authority: Is your site an authority on what the searcher is looking for? Having lots of high-quality, original content regarding the subject (in your case, information related to hospice and palliative care) signals that it is. So do incoming links from other legitimate sites Google considers authoritative on related subjects.

By filling your website with original content that informs and educates people on the process of palliative care, you're signaling to Google that you're an authority on the field — while simultaneously establishing *trust-worthiness* with human visitors that you'll take good care of their loved ones.

Social media: I'll discuss social media in much greater detail later in this book, but the more citations your site receives in social media, the more authoritative it appears on the subject.

Site performance: If your site receives lots of visitors, that's a big help. If those visitors not only stick around but also click on several other pages contained in the site, even better. Google tracks all these things, and they all affect your SEO.

You can view this as a cyclical process: The more visitors you get, the better your site will rank. The better your site ranks, the more visitors you get. So once you get that engine running, it virtually sustains itself. And your hospice website keeps growing in authority and popularity.

Content: *This doesn't just mean the words on the site.* This also means images, videos, and other media relevant to what the searcher is looking for. The more *original* the content you populate your site with, the better.

I like to say that ***content is king***, and for good reason: More than anything these days, the quality of your content will determine how well (or how poorly) your site ranks.

I talk more about content elsewhere in this book, but the key thing to remember is that your site *absolutely needs good content*. There's no shortcut around this that will produce the SEO and conversions you need.

I realize that very few independent hospices have the in-house resources to generate lots of strong content itself.

One option is to contract out the writing work. Another is to leverage the power of **artificial intelligence (AI)** tools such as ChatGPT to help you generate high-quality copy. These tools can produce a lot of copy very quickly and inexpensively, leaving you to spend your time running your hospice.

I'm going to dig into the pros and cons of AI tools later in this book, so please stay tuned for that.

Site design: Sites that are well designed are not only most helpful to human visitors, but to spiders as well.

A well-designed site makes it easy for the spiders to interpret how relevant the site is to what people are looking for. A properly designed site also includes keywords in its image tags and coding to communicate relevance.

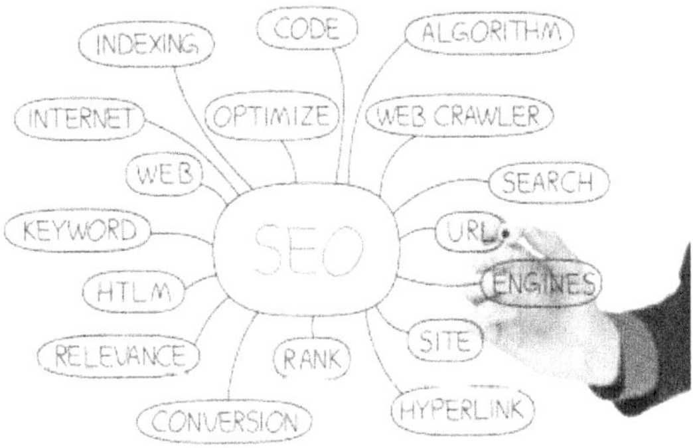

As for how to employ all these techniques and practices, that's a subject for a book that's much more specific to SEO.

You can use such resources to learn more about determining what keywords to use in your copy and coding, designing your site for maximum impact on human visitors and spiders, and tracking how well your site is doing compared to your competitors. Regularly analyzing and tweaking your approach is *absolutely necessary*.

If you prefer to concentrate fully on running your hospice, a professional internet marketing professional can handle all those elements for you.

Just remember to find a provider who uses white-hat tactics and can provide great examples of having increased sites' rankings through the legitimate methods detailed here.

It's all most beneficial to work with a company that has experience in hospice or related fields, because messaging and other key aspects are very different from other business types.

You don't want to contract with an agency that markets you the same way it markets hair salons or sandwich shops, of course. Marketing a hospice requires an understanding of nuance and thoughtfulness that's alien to many agencies.

Link building

As with SEO in general, link building is a broad subject. It's an essential part of SEO, because search engines still place a lot of importance on which sites your hospice website links *to* (**outgoing links**) and which sites link to *you* (**incoming links**).

If your hospice site is an island unto itself, that suggests to Google that your site is not very useful or influential, regardless of whether that's true.

That's why so much black-hat SEO is focused on artificially building up rankings based on link networks and purchased links. Not all link networks are necessarily "tricks," and not all purchased links are necessarily irrelevant, but they've been used so irresponsibly that it now requires great care to ensure you don't make Google think you're up to no good.

When you build links appropriately and organically, you're employing a powerful tool for your site. I'll go over a few of the basics here:

In essence, every live webpage in existence has the ability to "vote" for other webpages by linking to them. How many votes are provided by an incoming link depends on several factors, but a key one is the *SEO strength* of the site that's voting.

How much the votes from that incoming link affect your SEO also can depend on the *relevance of the linking website to yours*.

That's why getting inbound links from influential sites related to palliative care (or even healthcare or senior care in general) can be extremely useful.

It's also important to consider **link velocity**, which refers to how quickly your site acquires new links. If your site gets hundreds or even thousands of new links in one fell swoop, that's far less beneficial than steadily aggregating links over a matter of time.

If your hospice website suddenly gets lots of new links, that's a *huge red flag* for Google. If it steadily builds links over time, Google interprets that as a natural process for a site that's properly growing in influence.

The best way to get other sites to link to yours is to *create useful, interesting content* about the hospice field. A blog that's specific to your main focus is thus incredibly beneficial for attracting links.

Post regularly with original copy that's relevant to sites you want to link to yours. Make sure the copy is clean and grammatically correct: No one wants to link to a site that reads like it was written by Yoda after a three-day bender.

Also, don't only write copy that specifically sounds like you're pitching your hospice and its services. While it's fine to mention your business on occasion, you're far more likely to attract incoming links if you're providing solid information that's *useful for everyone*.

As I'll discuss further in the social media section, it's smart to blog about matters that are interesting to the demographic you're trying to reach, even if the subjects aren't exactly about hospice care. Your blog will generate a lot more readership and interest if it covers a broad spectrum of issues.

Also, adding images, infographics, and other types of content specific to your field makes your blog more SEO-attractive.

Again, you'll probably have to either leverage AI or contract out for content generation, unless you have someone on staff who's truly great at writing interesting, SEO-friendly copy (and whose time you can free up to work on it).

Other ways to encourage links:

- **Make it easy for people to bookmark your posts and forward them to friends.** Several tools let people do this with a single click. This is a great way to make posts go "viral" and gain links. (Remember, virtually any time of business — including a hospice — can create interesting content people will want to share. Open your mind about all the subjects you can cover.)

- **Include a link back to your website when you comment on other sites and blogs.** Most sites include a field for that, so use it. Obviously, be sure your comments are interesting and colorful, and that they represent your business in a professional manner. You never want a controversial comment — and *especially not a crude or profane comment* — linked to your site. That's true of any business, but obviously it's imperative in your line of work. If you delegate commenting you a staff member, make sure they know to always communicate in a thoughtful manner.

- **Submit your site to directories.** There are many different directories — some free, some requiring a paid membership — that will be included as incoming links when you get listed with them. Obviously, be sure to list your site with directories that are specific to the hospice field, along with your local chamber of commerce.

- **Link to other relevant sites whenever possible.** This refers to **reciprocal links**, which can be tricky. They aren't nearly as beneficial when they appear to search engines to simply be a one-for-one trade. (It's exactly like "I'll follow you if you follow me back" on Twitter.) However, in general, outgoing links help SEO when you're linking to quality sites, and sometimes those sites will link to you down the line. This will not hurt your SEO. Think of it as karma. As the Beatles said, "The love you get is equal to the love you give."

How to get in trouble: If you really feel like playing with fire and want to risk upsetting Google (and Bing/Yahoo), the best way is to obtain incoming and outgoing links by buying and selling them.

Search engines — and once again, I'm referring to Google in particular — truly *despise* the selling of links. It sets off all sorts of alarms that something bogus is going on. Buying links, while not quite as dangerous as selling them, also tends to backfire. Per Google:

The best way to get other sites to create high-quality, relevant links to yours is to create unique, relevant content that can naturally gain popularity in the Internet community. Creating good content pays off: Links are usually editorial votes given by choice, and the more useful content you have, the greater the chances someone else will find that content valuable to their readers and link to it.

As with SEO in general, there's more to generating great incoming links, but these are the basic tenets to keep in mind.

CHAPTER 2
Mobile Marketing

You'll recall that smartphones and tablets were mentioned often earlier in this book.

As of this writing (in fall 2023), we've reached the point where smartphones are truly omnipresent in the United States and all developed countries.

Today, they're easy to use, affordable, and incredibly convenient. Fifteen years ago, they might have been mostly limited to people in their wage-earning years, but now, teenagers (and even some adolescents) have their own smartphones. Grandparents and even great-grandparents have their own smartphones.

Of course they do. If you have teenagers, you want to know you can contact them anytime they're out of the house—and that they can contact you as needed. If you have parents in their seventies or eighties, you want them to be able to call you for assistance or whatever else they need.

Over the past decade, and especially in the last five years, smartphones have become a staple of everyday life for every generation.

If you were born in 2000, you have a smartphone. (Heck, you might have two.) If you were born in 1980, you have one. If you were born in 1960, you almost certainly have one.

Even if you were born in 1950, there's upwards of a 75 percent chance you own, and regularly use, a smartphone.

If we generalize to just Americans ages 50 to 64 — a key demographic for potential hospice patients and their families — a 2022 Pew Research Center study from Deloitte noted that over *85 percent* of that group owned smartphones.

That survey further noted that "smartphone growth was *strongest among older age groups* at over 10 percent" and that "Americans are viewing their smartphones more often than ever before, on average *52 times per day*."

Long story short: The demographic you're targeting has already made massive strides in smartphone adoption, and that's increasing every single day.

All of which makes perfect sense. Think about when you got your first cellphone. Suddenly, your landline didn't make a whole lot of sense anymore. You now had a phone in your hand that you could take anywhere. What use was it to have a phone that was wired to a wall back at your home?

If you still have a landline in your home, it likely serves as only as a complement to your mobile phone.

Similarly, the desktop computer is increasingly becoming only a complement to a laptop, smartphone, or tablet. Americans now access the internet *much more often on mobile technologies* as they do from home. And that's especially true of smartphones.

I don't have to tell you (but I will, because this is a book, after all) that kids today know how to operate an iPhone or Android phone before they even enter grade school.

Adults in their thirties and forties have been accustomed to communicating through mobile technology for much of their adult lives.

And it's very important not to dismiss the effect of mobile technology on older generations. A decade ago, you might not have expected people over 65 to be interested in carrying mobile phones. As of this writing, many have abandoned their landlines and use their mobile phones exclusively.

Mobile marketing takes many forms, but in essence, it's simply the process of marketing and promoting your hospice through mobile technologies (particularly smartphones), whether through your website, social media, text messaging, or other means.

This is where your audience exists now. While some still get their online information through desktop computers, mobile use has eclipsed desktop use for good.

Defining a broad term

The concept of mobile marketing preceded the explosion in smartphone and tablet use, but the term had a more limited definition, for the most part, previously.

For example, the first thing many people thought of regarding the term mobile marketing was text messages, more specifically **SMS**, which stands for short message service.

This later expanded to **MMS** (multimedia message service), which let companies text users who opted-in to the service with images, audio, and video. Both technologies remain aspects of many companies' mobile marketing campaigns.

Another aspect was the innovation of **QR Codes**, which let a user scan a 2D image with the phone's camera instead of typing in a URL or doing an online search.

I touched on QR Codes briefly before, but this is a good time for a reminder that this tech, which appeared to fall flat when it was first innovated, is now a part of everyday life.

Its massive resurrection occurred, in large part, thanks to restaurants having to remove physical menus during the thick of the pandemic. Suddenly, everyone—no matter if you were 18 or 80—had to get accustomed to pointing their smartphone cameras at these weird-looking codes to access a menu.

Now, you'll see QR Codes everywhere. At Hospice Haven Marketing, we use them in many of our print advertisements and mailings. It's just so easy for someone to point their camera at the code and tap the link that appears. Just like that, they're on our website (or landing page, depending on the objective)!

To be clear, we'll also include our website name. You don't want to limit potential clientele or referrers to just *one* way to reach you. Remember, a big key of marketing is to make it *as easy as possible* for people to learn about your services. (And we'll get into that more soon!)

Today, mobile marketing has taken on a much broader definition than it had in its infancy. It now encompasses the type of marketing and promotion you would do for PC users, but for people now accessing the internet on phones.

For example, many websites are built to display well on laptop or desktop computer monitors. You can view them in Google Chrome, Microsoft Edge, or Safari browsers to ensure they display correctly.

Those same sites, however, **display very differently when viewed on a smartphone.** They can be hard to read and hard to navigate on smaller displays.

This has led many businesses to create *mobile versions* of their sites. The site knows it's being accessed from a mobile phone and defaults to a mobile version.

Other businesses take this process a step further and create Android or iOS apps for their phones, letting the user simply tap an app to access functions of the business.

Today, it's more important that your hospice website be optimized for *mobile users* than for *desktop users*. That's not just an opinion: It's the way search engines such as Google do business.

Way back in 2018, Google officially implemented its **mobile-first index**. Here's the simplified explanation: When determining how highly to rank your hospice website, Google places much more weight on the mobile **user experience** (abbreviated as UX) than on the desktop experience.

In other words, if your site looks great and works well on a desktop computer but looks bad and is hard to navigate on a smartphone, Google *pushes you down in its rankings*.

If your site works wonderfully on a smartphone but not so well on the desktop, your site will rank better than the other way around.

Of course, I strongly recommend that your hospice website is optimized for *both* mobile and desktop users. For you, it's critical to have a great UX for both.

However, if you had no other choice but to choose one UX to optimize today, *it must be mobile*. That ensures that best rankings in search results and the best overall experience for both referrers and families.

(Final note: If we want to get granular, it's true that many new websites today have **responsive design**, which means they're created specifically to look great on all devices. Thus, if you have a newer site, you might be fine. But if you're not sure, you really need to view your site on a variety of devices right now.)

Mobile marketing is *the* key tool for local businesses

Mobile marketing is most critical for businesses such as hospices, who do all their business *locally*.

As I indicated above, it's time to stop thinking of your website or your online ads as being seen most often on a large monitor attached to a desktop computer.

Please forgive the bold type that follows, but I don't want you to miss this point:

Those days are over. Just like Google, it's time for you to think <u>mobile-first</u>.

Because you have a brick-and-mortar business that relies on local clientele, it's critical to be sure your marketing efforts not only include mobile, but make it a priority.

Some people will view your site on a desktop, sure, the same way some people were still listening to audio cassettes when the rest of the world was moving to compact discs, or how some still watched VHS tapes after DVDs arrived — and how a few stragglers are still listening to CDs and watching DVDs today, even though most of the world has moved past physical media and on to digital music and video and/or streaming.

Increasingly more people are viewing your site on a smartphone, and that will just keep growing. Mobile SEO and Local SEO are now *virtually synonymous*.

If your hospice's website isn't designed to automatically convert to a user-friendly mobile version when viewed on a phone, you'll lose virtually every potential client who pulls it up on a smartphone.

Not only does it take a lot more effort to read a full-featured site on a small screen, but you'll be seen by visitors as behind the times. If your local competitors have mobile-optimized sites and you don't, you're at a massive disadvantage.

Mobile devices also make it extremely easy to monitor statistics on who's checking out your site and listings, when and where they're viewing it, and which visitors chose to contact you or otherwise convert into a client. These analytics are critical to optimizing your hospice's marketing campaign.

CHAPTER 3
Social Media Marketing

Some people would argue the world was a better place when we could sit down in a restaurant and resist the urge to look down at our phones every 30 seconds to see whether they had a new notification on Facebook (or Instagram, or TikTok, etc.) or a new tweet on X (the recently rebranded name for Twitter, if it indeed sticks around very long).

Regardless, that's a reality these days. Some people say they're addicted to their phones, and it's hard to disagree. They communicate constantly through social media, and that communication, as noted earlier, increasingly occurs through smartphones.

The giants in social media — Facebook, X, and Instagram (and for younger people, TikTok) — wield astonishing power to let people and businesses interact with each other in real time. Having a presence on Facebook and X, at the very least, is a requirement for any online business.

A hospice, of course, is a little different. But it's not as different as you might *think* it is.

I'll elaborate on these points more in a bit, but here are a few things to know right from the top:

Yes, even a hospice can properly make use of social media to market its services.

Yes, there are types of content you can publish that are entirely appropriate for the services you provide.

Yes, social media can be an enormous influence on generating more admissions and referrals for your hospice.

I'll come back to those points in a bit. First, let's look at social media from a **return-on-investment** (ROI) perspective.

The biggest benefit to platforms such as Facebook and X is the cost: Their basic functions are free.

Sure, there are paid advertising functions that can be used as desired, and you'll see a lot of benefit from "boosting" posts on Facebook, in particular, but it costs nothing to set up a basic Facebook page and start posting.

The same is true of X. (Again, that's true at least of this writing. As of this writing, there have been tons of rapid changes at the former Twitter.) For now, X provides a means to reach out to potential clientele at no included cost.

Of course, money isn't the only valuable resource to a business. Time is just as valuable, and the time expended on social media is a factor in its use.

Your business might be set up in a way where you can easily communicate regularly on Facebook or X on a daily basis.

If it isn't, you might need to delegate those responsibilities to an employee or contract the process out to a third party, and those options require financial resources.

Whatever the case, every business should have a company page on Facebook at the very least. It's simply expected in today's digital world.

Having a Facebook page — and making sure to post on it regularly, even if it's only a short post or two every day — show that you're active and engaged in today's digital world.

Every time your hospice adds friends/followers on Facebook or X, you're adding to your firm's branding and influence. Every time you blog about subjects relevant to hospice and people comment on the posts or link to them, you're expanding your outreach.

However, there's one critical thing to understand whenever you use social media to market your business:

If you use it *unwisely* — being too focused on pitching your hospice's services or spamming followers with commercial links — you're risking a massive backlash.

There are few better ways to sink your social media marketing than by being too commercially oriented.

That might sound strange, given that your whole reason for social marketing is to increase awareness, traffic, clientele, and sales, but it's a quirk of social media.

People don't use social media to be "sold to." They use it to interact and to learn about things of interest to them. There's a lot of room *within that construct* to promote your firm, but for the most part, it has to be done *without a hard sell.*

(Honestly, I'm fairly confident anyone reading this book knows better than to take a hard-sell approach to supplying hospice services, but I figured.)

Use social media to make connections with interested parties. Share information, resources, advice, suggestions. You're a professional in the hospice field, so put that knowledge to use.

When people show interest in the services you provide, *that's* the time to provide information along those lines.

Social media can be an incredibly useful resource for business marketing when used wisely, but it can backfire quickly if you let the temptation of the hard sell get in the way.

Build relationships first; sell later.

Search is the best tool for leads; social media is best for referrals

This is one of the most important concepts to understand. In anything you do, you want to use the *right tool* for the job.

To increase the flow of leads and direct inquiries from family members, you need to focus on optimizing local search for your website. The better your site ranks, the more calls you get.

To expand your referral network, the best tool is social media.

This isn't an exact science, of course, but it's largely true. Social media is best at branding your hospice and creating a foundation for future referrals, but it won't generate a lot of direct calls for families. If they're searching for a hospice provider, they're not going to social media to find one!

By contrast, the more you optimize search for your firm, the more you'll attract calls from "problem-aware" clientele — that is, people who have an immediate need to solve.

In your field, that means referrers or families searching on Google or other search engines for local hospice providers.

While you could focus on one option or the other, if it's at all possible, you really want to do both. That way, you generate more calls with search and referrals with social media, *plus* you'll still get "transferable" benefits from both approaches.

Is a particular social network worth the time?

Not every social media platform is equal. Facebook remains the superstar of the social media world, particularly for the demographics you're targeting. (Other platforms might do well with X, Instagram, or TikTok, but the latter two are plainly not priorities for your field.)

Facebook is where you want to be. Although it has made algorithmic changes that massively limit organic reach (which means few people now actually see your "free" posts), it remains the best platform for targeting your audience.

(With that said, the best social media strategy doesn't exclusively target Facebook users. It's never good to put all your eggs in one basket.)

I've put together a few questions you should consider when determining whether a new social network will drive leads.

Who is using the network now? Not who the platform *wants* to use the network. Not who it *hopes* will use the network. Who is on there *currently*?

You should be able to identify the demographics of the network, how and when its users are interacting with it, why they are using it, and how involved they are.

How likely are these people to use or promote your service? Even if the people using the network fall into your target demographic, you need to determine whether they're likely to be involved with or interested in hospice services.

Are these the type of people using the platform? Is the platform itself set up to easily share items of interest between people using it?

The networks that provide the best return on investment tend to be the ones in which social sharing elements are integral to the platform and consistently used.

Facebook

You know your creation has become a landmark aspect of popular culture when someone makes a major feature film about both your creation and you.

The Social Network might not have always portrayed Facebook inventor Mark Zuckerberg in the kindest light, but the film's very existence showed what a gigantic influence Facebook is in people's everyday lives.

That's true even as newer social media platforms have stolen away some of Facebook's mojo. Younger people have fled Facebook over the past decade, virtually considering it "their parents' social media."

As quickly as things change, a future edition of this book in five or 10 years might be extolling the virtues of another social media platform and barely mentioning Facebook. But that's the future, and as of this writing, Facebook remains a major player.

Even as younger people have moved on from Facebook, the platform has *continued to grow in use overall.* That's right. Even as of fall 2023, Facebook is far from dying. It's still growing (albeit incrementally), now totaling over 3 billion users worldwide.

Facebook has become a constant companion in people's lives. And again, it's important to remember that Facebook is the more convenient, even the most *comfortable*, platform for middle-aged and older people.

It's incredibly easy to use. Mothers and grandmothers love it for keeping up with their families and distant friends. They're not giving it up anytime soon.

Advertising on Facebook: The social engagement aspect of Facebook is obviously useful, but again, it has its limitations, such as the need to avoid a hard-sell approach.

Facebook paid advertising is an entirely different animal. The platform provides tools that let you place ads that will conform to people's interests across Facebook.

This isn't a good place to get into the specifics, for two reasons. First, it can be a bit complicated. Second, Facebook revises its advertising approach constantly, so best practices today could be very different by next week.

Whether Facebook advertising will provide a good return on your investment is something to consider thoughtfully, preferably with the assistance of an internet marketing professional.

I can tell you that many clients have seen excellent ROI on thoughtfully crafted Facebook advertising campaigns. And at our agency, we innovate a number of different "recipes" that are designed specifically to provide excellent ROI for clients.

Boosted posts on Facebook

If you have any type of business, having some sort of presence on Facebook is a must. It's a largely simple and completely free way to build your brand, market your services, and communicate to clientele.

Actually, let's get back to that "completely free" thing for the moment. Yes, it's still *technically* free, but as with any very successful social media platform, Facebook is focused on monetizing its services.

If you use Facebook, you've probably noticed the ability to "boost" your posts, which costs money.

Boosting your posts lets you increase the time your post is seen by your friends, in addition to increasing the likelihood the post will be seen.

These days, Facebook really wants businesses to pay at least something for the promotion they get through the service. To that end, it changed its algorithm to massively depress the "organic reach" of posts on Facebook business pages.

In other words, the vast majority of unpaid posts you publish will *not* be seen in your followers' news feeds. It's unfortunate, but that's the deal. Recent studies show that organic reach today is often less than *2 percent*.

For this reason, it's smart to boost selected posts. Followers will see your hospice as an expert in this area and likely turn to you for these services.

Facebook Search: What you need to know

Facebook's search function pulls up relevant information based on you (the searcher) — your interests, your pastimes, what you care about.

In other words, your search results will be different from those of other searchers, because Facebook already knows what you're interested in, based on your Facebook activity and connections.

Facebook hopes that instead of simply using Google Search to find businesses and people of interest to you, you'll use its internal search function, which it believes will get you what you're looking for faster and more efficiently.

For your hospice, this means you should ensure you're active and well represented on Facebook, both on your own page and in interacting with users. That will increase the likelihood that people using Facebook Search will find your business.

Not only that, but you'll be prequalified by "Likes" from those users' friends and people within their fields of interest.

What's cool about Facebook Search is how it weights aspects that are similar to those of people (or businesses) to which you're connected on Facebook. For example, places that your connections "Like" or perform a check-in will show up higher on your search.

Facebook's search function makes the platform much more useful for marketing purposes, but as a Facebook user, you should also be aware that it makes information you might have considered private — or at least information you didn't necessarily want to broadcast to the world — far more easily accessible.

If you've "Liked" certain pages without much consideration, be advised that Facebook Search can show all that info *unless you've specifically restricted that information* in your privacy settings.

Similarly, if other people have restricted that information in their own settings, you won't get "hits" for them when you perform a search.

As you're likely aware, there's been a lot in the news the past couple of years about Facebook and privacy, in particular the ways it's been monetizing people's data, among other concerns.

If you're worried about that, you're perfectly right to be. If you're especially concerned about your personal data (on your personal Facebook page), you should use Facebook's privacy settings to limit the personal information you share.

You can even quit the platform altogether (again, as an individual) if you're truly apprehensive. I get it.

However, this book is about growing your hospice and how Facebook is a powerful tool for doing that. So if you have concerns about Facebook in general, I suggest weighing those against the great benefits it can provide your hospice.

To that end, you should use the Facebook Search function to make sure everything you want users to find when they search for local hospice services will show up. That means getting your hospice's social profile in order.

Why? Because these social connections are pretty much the "links" that Facebook uses to determine relevancy and authority.

It's these photos, likes, check-ins, recommendations, etc. that the Facebook Search algorithm will use to qualify and personalize results for Facebook users.

In a very real sense, "Likes" are becoming the "links" of the new digital era on Facebook. What you Like defines you (and/or your business). The more you Like hospice-relevant posts and pages, the better you'll do in Facebook Search.

X (formerly Twitter—at least for now)

Honestly, it's weird for me to keep referring to "X," because we've all known it as Twitter for so long. And its current stewardship (again, as of this writing in fall 2023) has been chaotic, to say the least.

I have no idea what this social platform will look like down the line. But as of this writing, its owner is calling it X, so when I refer to it in the present tense, it's X. (Just please try to remember that it's what you used to call Twitter.)

When then-Twitter arrived on the scene, business owners were skeptical about how useful this tool could be. If all you can do is post something in 140 characters, how useful can that be?

In fact, it turned out to be extremely useful, especially once it added options to let you write longer posts (with the remainder represented by a link), to include link shorteners (so your characters weren't all used up by a huge link) promoted accounts and posts, etc.

In 2018, then-Twitter finally expanded the character count for basic tweets themselves, letting users post up to 280 characters. For reference, the very paragraph you're reading right now (including spaces) is 246 characters. That's plenty of room!

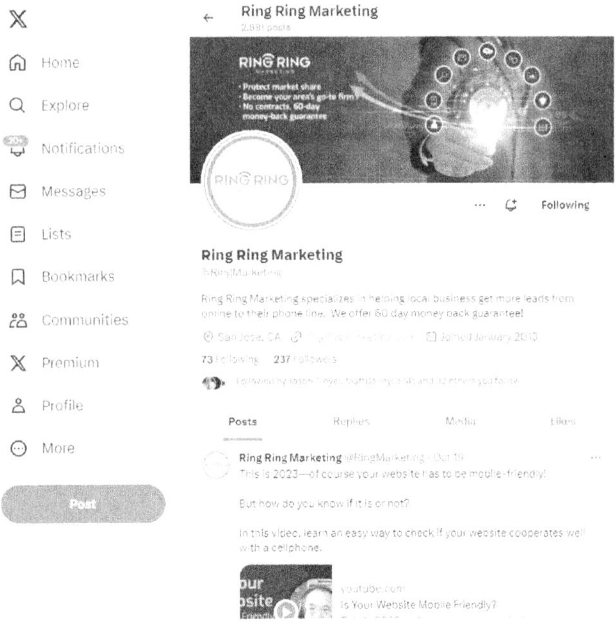

More recent developments allow verified users (who pay X a modest fee for the privilege) to provide much-longer posts, turning this onetime microblogging site into, well, just a regular blogging site.

With that said, I can't tell you that's still the case if you're reading this in fall 2024 or anytime in 2025, 2026, etc. Like I said, it's been really chaotic over here.

For that reason, I won't dig into X too deeply in this book. When it was Twitter, we largely knew what to expect. At the time I'm writing this, its business model changes practically daily.

Yelp

While it's certainly true that few people searching for hospice services will prioritize searching on Yelp, it does remain an influential review site.

Yelp was one of the first platforms to mix user reviews with a social media aspect, and while it's true that Google reviews remain far more important, Yelp does matter.

As of summer 2023, Yelp was the second-most used review platform in U.S. market share. However, that's quite deceiving: Yelp's runner-up spot represented just *6 percent* of the market share. First place went to Google, with *73 percent* of the market share.

However, as I noted earlier, Bing aggregates reviews from outside platforms, including Yelp. So if someone searches for a hospice on Bing—which is often true of people who haven't changed their default browser and search options when they buy a new Microsoft computer—those Bing reviews could pop up.

Blogging

While this is another area that might sound uncommon for the hospice field, the benefits of having your own blog are substantial.

Blogging about hospice-related issues in a way that displays expertise provides you enormous credibility, which goes a long way, especially for potential physician referrals. It lets you communicate about key issues. It lets people provide feedback and ask questions through comments.

(And you can moderate those comments before they're posted, removing the risk of spam or other problems.)

Then there's the SEO consideration: When you're generating a lot of content for your website that's relevant to palliative care, your SEO increases *massively*.

Having a strong, regularly updated blog is one of the very best things you can do to increase your hospice website's traffic and search engine rankings.

Blogging also helps you create incoming and outgoing links, which are great for SEO. You can (and should) link to other relevant sites and articles in your blog posts. When you post something of interest to others, it's likely your post will be linked to (or shared through other platforms), increasing your incoming links and traffic to your site.

With all that said, there are some obvious potential downsides to blogging. Spelling, grammar, or fact errors in your posts can make you look unprofessional.

So if you're not a good writer (or sufficiently detail-oriented), you should have a better-suited employee or a contracted professional do your blogging for you. Alternatively, you now can employ AI content creation to generate posts that fit the bill (as I'll elaborate on later).

Write original content whenever possible. Posts don't necessarily have to be long; even a couple of paragraphs can be enough, as long as the content is focused on something that interests your target demographic.

As with everything else, blogging is more beneficial for certain types of businesses and services than others, but it has some benefits no matter what you do. Just its ability to improve your site's SEO alone is a good reason to try it.

If this sounds like too great a time investment, speak with a marketing consultant about possible options to outsource the process.

A lot of people make their livings blogging for a variety of different sites, and depending on your needs, these services can be very affordable, especially given the return on investment.

LinkedIn

A social network specifically for business professionals: a simple enough concept, and one that's worked out for LinkedIn to the tune of more than *900 million* members. (That's 400 million more than it had in 2019.)

Don't get too hung up the term "business professionals." You don't need to wear a suit and tie to work to benefit from LinkedIn. That's how pervasive it has become for business networking.

LinkedIn simply makes it incredibly easy to allow registered users to maintain a list of contact details for people with whom they have some level of relationship. These people are called **connections**.

Once you've made connections, you can connect with people connected to them, allowing you to easily expand your networking base with people in your industry or with common interests/work histories.

You can create a comprehensive profile with a full resume and qualifications, and when you're looking for a new employee or business partner, you can search through LinkedIn users for the experience, skills, and qualifications you're looking for.

LinkedIn also lets you endorse people you've worked with — and be endorsed by the same — with the click of a button. Having your business and various skills endorsed by well-connected users is obviously a great way to gain new business and new opportunities.

Here's the usual caveat: LinkedIn is most beneficial for businesses in which professional networking is an integral part of doing business. However, the platform has a lot of additional tools that can be useful for virtually every type of business, including a hospice.

LinkedIn doesn't require constant participation, although the more you participate on the platform, the more helpful it can be.

For a hospice, linking to others in the field and those who provide related services (albeit not your competitors) can provide a great foundation for organically building incoming links.

At the risk of stating the obvious, connecting on LinkedIn with physicians who provide elder and/or palliative care, along with nursing home and assisted living administrators, could be a massive benefit for referrals.

When their sites link to yours, it improves your traffic and your organic SEO, because Google values incoming links from entities who are also in your field very highly.

Artificial Intelligence (AI) and Content Creation

While we're on the topic of social media, let's explore how you can find time to create content for social networks—and website copy, and promotional materials, and so on.

One way is to leverage the power of *artificial intelligence.*

While artificial intelligence (AI) certainly became one of the most broadly discussed (and deployed!) technologies in 2023, it isn't a new concept. Here's a little history:

In the 1950s, Alan Turing, a British mathematician, proposed the Turing Test, a method to ascertain if a machine could exhibit intelligent behavior indistinguishable from that of a human.

Around the same time, John McCarthy coined the term "artificial intelligence." In 1956, McCarthy organized the Dartmouth Workshop, marking the birth of AI as an academic discipline.

From the late 1950s to the 1970s, the field experienced its first "golden age". Early successes, such as the General Problem Solver and the development of the first neural networks, sparked enthusiasm and investment. Popular media began portraying AI with intrigue, often casting it as a powerful, sometimes ominous force, as seen in movies like *2001: A Space Odyssey* (1968).

However, the complexities of AI quickly became apparent, leading to reduced funding and interest, known as the "AI winter" periods. These cycles of boom and bust continued through the 1980s and 1990s.

The 21st century heralded a renaissance for AI, primarily driven by advancements in machine learning, notably deep learning. The availability of vast datasets and powerful computing resources enabled neural networks, inspired by the human brain, to flourish.

This reawakening led to breakthroughs in various applications: from image and speech recognition to game-playing AIs that could defeat world champions, like AlphaGo's victory in the game of Go.

Popular media has played a pivotal role in shaping the public's perception of AI. From the benign and helpful robots like R2-D2 in *Star Wars* to the malevolent Skynet in the *Terminator* series, AI's portrayal has spanned the gamut of emotions.

Movies, literature, and television have often mirrored society's hopes and anxieties regarding AI. While films like *Ex Machina* and *Blade Runner* have raised philosophical questions about consciousness, identity, and the rights of artificial beings, others like *Her* explore the complexities of human relationships with AI entities.

Let me take a moment to interrupt the history lesson, because here's something I should acknowledge before we go any further:

If you read back a few paragraphs to "In the 1950s…" through "AI entities," **I didn't write any of that.** Really! Outside of a few tiny tweaks, all of it came directly from ChatGPT-4, one of the most popular AI content creators at the time of this writing.

Did you notice the difference? It's fine if you did. ChatGPT's style isn't exactly the same as mine. The latter is a little more straightforward and "textbook" sounding. But the information is still quite accurate and well communicated.

Artificial intelligence and small business

Here's why we're discussing AI in the first place: As of 2023, certain AI tools, particularly "chatbots" such as ChatGPT, and Bard, have become incredibly user-friendly for virtually anyone — including small businesses such as yours.

That's the short version. The longer version is that AI tools such as these come with various pros and cons. You've probably heard of some of these.

Some people believe AI is going to eliminate the human workforce in a variety of fields or consider chatbot content tools "plagiarists" because they're developed through machine learning of original content created by (you guessed it) humans.

Others believe these tools promise incredible benefits for human workers who know how to use them wisely, merging the content chatbots deliver with irreplaceable human knowledge.

The truth, frankly, is somewhere in the middle. There are dangers to AI. And there are amazing benefits. Ultimately, it comes down to using this technology wisely and understanding both its promise and its limitations.

Let's just look at ChatGPT for a moment. Once you understand some (very simple) prompt engineering, you can get ChatGPT to create high-quality blog posts, social media, marketing messages, promotional content, whitepapers, and much more. (And that's just the tip of the iceberg, one small portion of what it can do.)

With that said, ChatGPT is just a tool. And any tool is only as good as the person using it. ChatGPT doesn't have an actual brain capable of independent thought. However, it was programmed (and continues to be) by some of the smartest people on the planet, people who know how to provide the best possible content given the fact that you are, at the end of the day, just talking to an incredibly complex robot.

In other words, ChatGPT and a Roomba aren't all that different at the end of the day. They can only do what they're programmed to do. They only perform based upon what they've been taught. It's just that one vacuums your floor and the other one, well, doesn't. (But it does a heck of a lot more.)

The truth is that you likely use AI every day. It powers the search engines you use to learn information online. Every time you ask Siri or Google or Alexa for directions, recipes, or a piece of trivia, AI creates that response. When Spotify recommends songs based on your listening history, that's AI. (And again, those are only a few small examples.)

I could go on about AI all day, but I know your time is important, so let's focus on AI's current and future benefits for hospice care. Let's further limit it to the ramifications for marketing your business, because that's what we're here for, after all.

I can tell you that Hospice Haven Marketing regularly uses AI in a variety of ways. We use chatbots to help create informational guides, social media posts, emails, lead forms, and other content.

To be clear, this isn't a "set it and forget it" process. Our staff has taken extensive training on how to create high-performing prompts (the queries you create for an AI chatbot) to get the results we need. And information the chatbot creates for us is then carefully scrutinized, and edited as needed, to ensure excellence in everything we deliver.

However, it's honestly not that hard to do. A little study of how to use chatbots can get you up and running in no time. And even if you're not the strongest writer or editor, what you get back from a well-crafted prompt will be better than what most people can produce without any AI help.

I'll grant that I understand the concerns of skeptics and those who think we're on the edge of a slippery slope. For example, many teachers and professors have railed against the use of AI by students to create essays and term papers.

As I'm writing this (in fall 2023), Hollywood writers just settled a strike where use of AI by studios was a major concern, and the actors' guild (again, at this time) remains on strike, in part for similar reasons.

Those are fair points. However, it's also true that AI developers and other engineers do provide fairly robust tools for checking whether something has been created by AI (important for academia), and we're still not a point where AI use in movies and television is an existential threat to human creators.

Another oft-voiced concern comes from visual artists who note that whole libraries of their creative output have been used in machine learning for AI-based art applications. They consider that plagiarism. And there's some validity to that.

But none of that changes the fact that user-friendly access to incredibly powerful AI tools is here now, and it's very likely here to stay. Some regulations and controls might alter the landscape in time, but it's simply too useful and powerful to ignore. And frankly, that's a good thing for businesses such as yours.

I'll never say that human creatives aren't important. I have many highly skilled, extremely experienced people on my staff overseeing and monitoring the development of all our creative content, whether it's to promote our clients or HHM itself. The human factor must always be a part of any AI initiative.

But it's also true that AI helps those very people create the very best marketing and promotional material you'll find. Again, *AI is a tool*. Spellcheck helps us all write and edit better. That's AI. Word processing software itself is a tool. Graphic design software is a tool. AI is just an extra layer to help humans get the most out of their creativity.

So if you've been thinking you'll never have time to add content to your hospice website, think again. If you've wanted to post on social media to strengthen your branding and attract more referrals, here's a solution. If you've wanted to do email marketing or create better physical handouts promoting your hospice, now you're all set.

These AI tools won't do all the work for you, but once you jump in, you'll be amazed and how much easier and faster it can be. If you want to work with a marketing firm such as ours, we have all the processes in place, but if you want to DIY it, quite honestly, you can do a pretty good job on your own!

And even if you don't love the idea of a DIY chatbot actually creating copy or images for you, tools such as ChatGPT can be incredible idea generators. If you want to come up with 20 potential topics for blogs or social media posts about hospice, just ask the tool. It will provide lots of ideas, many of which likely will fit the bill perfectly. (Others might not, but that's a given in a process like this.)

The ultimate lesson is that technology advances quickly, and AI chatbots represent just one more evolution in the process. If you want to have a competitive advantage in business, you need to be open to these advancements.

Section 2:

Conversions

CHAPTER 4
Maximizing Website Conversions

In the digital age, just as internet marketing essentially *is* marketing, your website *is* your hospice's home.

It's as crucial as having an office. Your hospice could change its physical location, but its website is a constant, the place where anyone can find you in this digital age.

Today, where you and your business are physically located matters, but only so much. Where you exist *online* matters a great deal.

Perhaps you're thinking: So what? *Of course* my hospice has a website. I pay a few dollars every year or two to the domain provider and hosting company.

My phone number is on there, along with a picture of the business, maybe even a few paragraphs about what we do. I've done my due diligence.

Well, that's just the bare minimum. That's like putting on shoes because the grocery store won't serve you otherwise.

Simply having a website is like simply having a business. It exists. Now what are you going to *do* with it?

Like every other aspect of your firm, your website has a purpose. We can refer to that purpose in general ways, but it all ultimately comes down to one thing: *conversions*.

I touched on conversions earlier, but let's dig in a little deeper:

People define "conversions" many different ways. For businesses where the goal is online sales, a conversion is a very simple thing: a sale. It's *converted* a visitor into a paying client.

As the operator of a hospice, you earn revenue through providing care services. These things take place in person. So it's not as easy to gauge how effective your website has been in converting visitors by using basic sales analytics.

However, there are many analytical methods to determine how effective your website is at attracting new visitors and determining how they got there, how long they spend on the site, what pages they click on, and how long they spend on those pages.

These tools let you see which images, videos, and links they click on. And they let you know what's working — and what's not working — in attracting clientele.

I've said it before, and I'll say it again: People are now finding services (and products) online far more than ever before. If they want to find a dog walker, a bike shop, or a psychic, they're finding them *online*.

Businesses that have no website at all? They're in massive trouble. To the ever-growing digital population, those businesses don't exist.

Businesses that have a bare-bones website? They're almost as invisible as those with *no* website. With little-to-no content, no SEO to attract visitors, no incoming links to provide inroads for potential clients, a website like that is a needle in the proverbial haystack.

Businesses with a middle-of-the-road website? That's 100 times better, no question about it. If it has some content, some degree of SEO, and uses some links — and if that's tied in with a touch of social media — that website is actually achieving the minimum standard of what's needed to be successful today.

However, you can do better. The jump from a bare-bones website to a middle-of-the-road one is huge, no doubt. It's like going from a Burger King to an Outback Steakhouse.

It's the next step that makes all the difference, the one that jumps from a chain steakhouse — a perfectly serviceable chain steakhouse, but a chain steakhouse all the same — to a five-star restaurant.

That's the step to a website that's actually oriented toward *converting visitors into referrers and clientele*.

And yes, it can still be designed in a way that's extremely tasteful, caring, and respectful, which is imperative in your field.

This is one of the most powerful revenue generators in modern business.

And obtaining this goal is a lot easier than opening a five-star restaurant. (There's certainly no need to hire a fancy executive chef.) By employing a set of very basic tools, you can turn your website into this powerful revenue generator.

If you're going to have a website, you might as well put it to work for you. That's what it's there for. If you're not maximizing its potential, you're leaving an incredible amount of money on the table.

Don't let that happen. Get your website to not only attract visitors, but to convert them into referrers and clientele.

Designing your website for conversions

Here's the first thing to understand: No visitor will convert until your site has proven itself. Every element of the site must underline your professionalism and credibility.

It must say to the visitor: *We're the right choice for you. There's no need to look elsewhere.*

If you can convince the visitor that you're the best choice right then and there, you've removed any need for the visitor to leave and keep checking around. *The visitor wants you to be the solution.* You just have to make it happen.

Only after you have proven yourself to your visitors can you focus on converting visitors to clientele.

Use great testimonials: You have many satisfied clients, right? Families you assisted during an exceptionally challenging and overwhelming time? Ask a few to write a testimonial.

Visitors want to know that real people have been incredibly impressed by the professionalism and care you showed when they've dealt with your hospice. Trust is obviously a powerful consideration in this field.

You can include more testimonials elsewhere — devote an entire page to them if you like — but use the best ones on all pages where you're trying to compel the visitor to contact you regarding a referral or services.

Of course, testimonials are effectively the same as online reviews. I'll get into reviews and reputation more later, but you should know that the more great reviews you receive, the better you'll convert website visitors.

Address your visitors personally: The personal touch is very important, especially in your field. People want to know you're a real person who is proud of your hospice.

Put together some copy to tell your visitors who you are, how important hospice care is to you, and why you and your staff are the best people to help.

Show why the visitor should work with you: Here's where you explain why your hospice is the best option.

Think about why someone should choose your hospice over anyone else. It's like writing a resume or a cover letter when you want a job: The employer has a problem (a job opening) that needs filled, so you're explaining why you're the best person to solve that problem.

Just as in a resume, don't get too bogged down in details: Focus on the most important facts that sell your hospice on the visitor. Whenever possible, include as many *quantifiable* facts as you can.

This can be a hard nut to crack, but it's absolutely worth the time. Brainstorm with your staff and really figure out why you're the best option for helping patients and families in need. What sets you apart? What is your differentiator, your **unique selling proposition** (USP)?

Once you know this, you can drive that point home on your hospice website. Know your USP and really impress it upon the visitor. That's one of the best ways to drive conversions.

And here's another one:

Call to action: There are some other copy elements that can help reinforce your pitch, but they can vary depending on a number of factors, so let's cut to this integral aspect:

Your site needs a very clear **call to action** (CTA).

Most often for a hospice, that call to action will be soliciting the visitor to call your telephone number. If the visitor does so through a click on your website, it will be convenient *and* trackable.

The CTA could be to click on your email address link (it *must* be clickable — *always* make it easy for the visitor), which is another thing you can track.

The CTA also could be to fill out an information form, which could benefit families who are planning ahead with the expectation of needing hospice services in the near future. (Keep this form short and sweet; that's what you'd want when you visit a site.)

Impress upon the visitor that you're the best option for a referral or care services, and your help is just a phone call away.

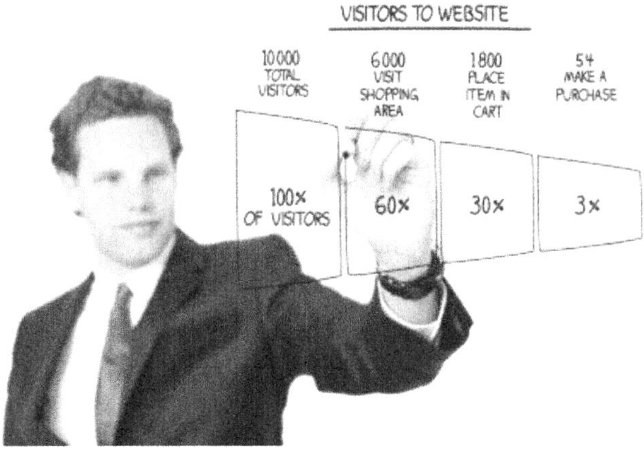

Improving your calls to action

The CTA is one of the most important aspects of your website or landing page. You attracted a visitor to your site for one reason: to act in some way. Here are a few ways to improve your calls to action:

Include the CTA on every page on your site: You never want people to have to search for your hospice's phone number or email address. Make it easy. If your website is designed so that the CTA is a stable element throughout the site, the visitor can interact with it at any point, on any page.

Specifically, this means you want your web designer to create a **fixed header** (aka a **sticky header**) that stays put throughout the navigation experience. If the user scrolls down on a particular page, this header remains in place. If the user navigates to another page, the header remains.

In addition to ensuring the CTA is clearly displayed throughout your hospice website, this creates a uniform look and feel. No matter where the visitor travels, your business name and contact information are boldly displayed at the top.

Make sure your CTA stands out: You obviously never want anything on your hospice website to be obnoxious or "salesy," but make sure your CTA stands apart. And yes, it can be polite and respectful but still be distinctive from everything around it.

For text CTAs, use a different color for hyperlinks so it's clear to readers that the text is clickable and an action can be taken. For visual CTAs, use a color that starkly contrasts with the rest of your page.

Use strong, active language: This isn't the place to be passive or longwinded. Everything is funneling toward the visitor taking an action, so make the CTA clear, simple, and direct.

The visitor should understand exactly what he or she is getting and why acting on the CTA (whether filling out a contact form or calling/emailing you) is the logical thing to do. If that isn't being clearly conveyed, make a change.

Be sure the value of acting is obvious: If you can't explain to a visitor why your offer is going to help in some way, why would the person click? Your visitor came to the site to gather information about hospice services: The CTA needs to be the solution.

A brief overview of contact fields

According to recent surveys, most marketers consider the lead-capture form to have a very significant impact on website performance.

You have a very short window of time to capture anyone's attention when that person visits your page — many experts put this window at a mere *six seconds*.

This rule is something to consider when designing your lead generation form. If the form seems too arduous or time-consuming to fill out, most visitors won't even bother.

The key is finding a balance between getting the information you need and keeping the form simple enough that it won't scare off your visitor.

Some studies indicate that indicated that the optimal number of fields to include is seven, but I don't agree. Keep it simpler than that.

Get a *name*, *phone number*, and *email address*. Just three fields. If you really want to ask for a physical address, I understand, but including that field *could* scare off the lead.

I prefer the strategy of getting basic info at first, "warming up" the lead through marketing emails and related techniques, and then working to obtain a physical address.

If you include too many fields, you'll see a significant drop-off in conversions. If you include too few, you're in danger of missing out on valuable information you might need.

Whatever you do, don't get into the double digits: That's as good as telling visitors you don't consider their time valuable. They're providing you information, not taking the SAT.

Keep in mind that there's a big difference between what information you'd *like* to have and what you *need* to have.

Name, email address, and phone number are essentials. For a hospice, you *might* ask for a ZIP code and/or and some other demographic information, such as household income or age range.

But again, the more fields you include, the fewer conversions you're likely to get.

Ultimately, one of the best things you can do is review your landing page as if you were simply a visitor: Would you want to fill out that form? What advantage will you get from providing your contact info?

If your responses aren't overwhelmingly positive, you'll want to make some changes.

Grabbing your website visitor's attention right from the top

Visitors are great, right? Well, maybe not visitors to your home, at least when they're annoying in-laws or someone selling appliances door to door. But getting visitors to your hospice website is great.

If they're coming to your site, you're popular. You can check the analytics and see where they're coming from and what brought them there.

Of course… visiting the site is only half of the process, and you don't get any credit for half. A half-baked cake isn't something you want to eat. A half-vacuumed carpet just looks worse than a carpet that hasn't been vacuumed at all. You know what they say: Close only counts in horseshoes.

As I noted previously, visitors do you no good if you can't convert them into clientele.

It's like having the most-visited art gallery in town, but if no one ever buys a painting, you'll be the most popular artist begging for quarters on the street.

Your hospice website can't just be focused on attracting visitors; it needs to convert them into referrals and (ultimately) admissions. It needs to get the point across right away.

How do you do that on your site? For one thing, figure out your top "selling" point to potential referrers or clientele and try to make that the first thing they see. (My apologies for using terms such as "selling," which I know sound irregular in hospice, but that is simply the proper term for this context.)

Your hospice's name is all well and good, but that alone doesn't sell the visitor. Use that valuable real estate at the top of your site to sell your visitors on what your hospice does better than anyone else: your *USP*.

Granted, a hospice website should always be respectful and not have a "hard sell" feel, but there are definite ways to impress your qualities upon the visitor without seeming crass.

Whatever you do best, whatever your claim to fame, spell it out clearly and make sure visitors can't miss it.

A note about SEO: As I noted earlier in this book, an integral aspect of ranking highly in search engine results is incorporating relevant keywords in your copy.

These need to be in there, but far too many sites use keywords incorrectly. They stuff copy full of keywords to the point where the writing sounds like it's intended for the spiders (the search engine software that crawls the site), not for the visitor.

Never do this.

It will send visitors racing for the "exit." It looks especially amateurish on a hospice website, but it honestly looks bad no matter what business you're in.

A qualified internet marketing expert can help you learn how to incorporate keywords so they sound natural in the copy.

Also, when keywords are used *too* often, this can actually *hurt* your SEO. Counterintuitive as it might seem, clunkily jamming in "hospice" or "palliative care" into every single paragraph works *against* you.

The search engines will penalize your site in the rankings if the keywords appear too often, or not in the right places.

Yes, it's important to serve two masters, the human visitor and the search engine spiders, but always err on the side of writing intelligent, compelling copy for the visitor.

Nothing will undermine your credibility faster than copy that makes little-to-no sense because it's only there to drive search engine rankings.

Conversion optimization factors

Let's go over some additional factors that help ensure conversions on your website:

Structure your navigation: On too many websites, the site navigation is cluttered and disorganized; it doesn't help drive potential clientele to the pages that will give them the information they need. One of the biggest mistakes is creating site navigation that links to virtually every page on the site.

When you provide so many options, it's confusing to the visitor, and you're not taking advantage of the opportunity to lead the visitor through a page-by-page process to deliver the message you want to convey. It's like suggesting a reader jump around randomly from chapter to chapter in a novel.

No matter your site's size, try to pare your navigation down to as few elements as possible. It's okay to have a dozen links or so in your navigation, but be sure they are *organized* in a way that helps visitors easily choose the right page.

Implementing structured navigation makes it much easier for visitors to quickly find the pages that are most interesting to them.

A hospice website doesn't have to be overly complicated. You want to make sure there's good, detailed content on every one of the pages, but that doesn't mean you need a *lot* of pages.

The easier you make it on the visitor, the more likely you are to get a conversion. Carefully scrutinize every part of your site and see where you can simplify the process.

Make sure your site has a voice: If your copy sounds like every other hospice, you're missing an opportunity to differentiate your business from the competition.

While the tone of your website's copy should always be professional, "professional" doesn't have to mean "antiseptic."

Too many people believe that business-oriented copy needs to read like the most academic of research papers. *This is a huge mistake.*

Copy can be professional but still sound *resonant* and *personal*—even on a hospice website. (In fact, it's particularly resonant in this field.)

Read through your copy and see if it sounds like you're simply having a thoughtful conversation with a potential patient or family. That's the tone you want. The more personal flair you can inject in your website copy, the more it will stand out from the vanilla sound of so many other sites.

Provide a box that lets visitors search your site: This depends on how well your hospice website has been developed, but if you have good, useful copy on all your pages, a "search this site" feature can be a great boon.

If you're not sure how well your site delivers results when people search for certain common terms or phrases, test its functionality yourself before letting the feature go live. A search function that works poorly is worse than no search function at all.

If you do have a great site search that gives perfect results every time, this can help visitors find specific products, features, or information quickly without having to scan a bunch of pages.

This search data can also be helpful in telling you what people are looking for, allowing you to better promote those areas.

Be consistent in formatting (fonts, bold, italics, images): There should be a consistent look and feel to all your pages, making them seem (as they are) all part of the same message.

While some homepages may look a bit different than the internal pages, it is important to be consistent from one page to the next. Your primary navigation should not change from one page to the next, nor should your footer, page layout, colors, etc.

When you don't have that kind of consistency, the onsite experience can be quite jarring, and this can ruin the experience for visitors. Any bad experience is bad for conversions.

Make sure your site is optimized for mobile: I covered this extensively earlier, but I'm just noting it again here because it's absolutely *essential* for conversions.

While hospice sites still get more desktop traffic than most other industries do, you'll still be getting well over half of your traffic from mobile devices.

No one using a smartphone wants to deal with an old-school website that isn't optimized for mobile. They want to call you with a single click. They want to be able to look over your services in a clear, easily viewed format.

If you're not optimized for mobile, you could be throwing away more than half of your potential conversions.

Delivering great conversion copy

I hit upon a number of content considerations earlier in this section, but let's talk a bit more about persuasive copy and how it can convert visitors into clients.

More isn't always better: Don't overload your pages with thousands of words of copy. You're trying to deliver a message, so provide just enough information to get that message across, and then get out of the way. You don't want to bury the visitor with minutiae that don't build your brand or deliver a conversion.

However... don't take that to mean that a couple of paragraphs on each page will get the job done. Your copy needs to be detailed enough that people clearly understand what you're conveying and accept you as your area's top expert on hospice care.

If you have a Frequently Asked Questions (FAQ) page, be through, but resist the urge to pack it with tons of salesy language. Provide honest-to-goodness answers to actual likely questions.

For the most part, keep sentences short and to the point. Don't repeat information that you've covered elsewhere on the site.

Your visitor's time is valuable, and digging through long stretches of copy to find necessary information will cause a visitor to look elsewhere. Everything is a balancing act: Provide strong, detailed, concise information, and then wrap it up.

Did I make it perfectly clear that you shouldn't skimp on content? Content really is king, and the higher quantity of strong content you have on your hospice website and your pages, the higher the traffic and conversions, generally.

Content-rich sites also attract more links, which is always a good thing (presuming the links are legitimate and relevant). Be sure to have plenty of content; just make sure that the content you have is broken up into *clear, concise, easily digestible* chunks.

Show that you're trustworthy: Why don't people convert even though a site clearly shows that it can provide a definite solution to the visitor's need?

Usually, the biggest roadblock is that the site hasn't *sufficiently communicated its trustworthiness* to the visitor. I'm sure it's rare that you get a referrer or client who picked a name out of a hat for something as important as hospice services.

Because far too many bad businesspeople have promised too much and delivered too little in the digital age, online consumers tend to have their shields up.

Your copy should be consistently focused on putting a potential client's mind at ease. Testimonials (both written and especially in video, as noted earlier) can help with this.

The visitor wants to believe you. A potential referrer wants to know they can feel confident recommending you. A family or patient wants to know you will care for them well. As the owner, you have to conquer any concern in the visitor's mind that you might not be the best choice.

Trustworthiness, transparency, credible authority, lots of high-value content, and plain old decency: These are your best weapons.

These factors would be true of any business, but they're obviously critical for a hospice.

People might buy some gizmo online if it's really cheap, even if they're not 100 percent sure the seller is entirely trustworthy. They might grab a burrito from a random food truck at the park if they're famished, holding out hope that it won't make them sick.

But it's another thing to ask them to work with a hospice they don't trust. It's your job to make that perfectly clear through the information you include on your website.

Everything on your website needs to show you can be trusted. Provide real contact information. Display your photograph. Include pictures of your hospice. Include a short video featuring you and your staff. Make visitors feel like they've already visited your location before they've even set foot in it.

If you're a Better Business Bureau member, make note of that (and display the seal). If you're affiliated with national or state hospice organizations, put their seals up there. Display positive reviews and testimonials. That's what we call **social proof**. It's a key aspect of ensuring more website conversions.

Trust, trust, trust. That's what you're selling.

Check the readability of your copy on various browsers, including mobile browsers: Nothing will run off a visitor faster than a website that's hard to read. The fonts and colors you use might look wonderful in Google Chrome but awful in Mozilla Firefox or Microsoft Edge.

Test the readability of your site in all types of browsers and select options that display clearly (and in a compelling way) across the board.

And of course, make sure the site is optimized for mobile — and that means it displays and functions well on both iPhones and Android devices.

While SEO is important, always put the human visitor first: Yes, I'll repeat myself one more time: If your copy reads like it's intended for search engine spiders and not the reader, *you're not going to convert*. This practice massively undermines your credibility and serves to confuse the reader. Get your keywords in there but be certain that the copy is personable and conversational in tone.

Finding the balance in copy that's compelling for readers yet also keyword-rich is an art of its own, and a good internet marketing expert and copywriter can be a great help with this.

Onsite organic SEO

Effective SEO isn't simply a matter of getting your site to rank as highly as possible in SERPs or driving as much traffic as possible to your site.

It delivers the *right* people to your site, people who are likely to convert. To achieve this, you want to use relevant keyword phrases not only in the copy people read, but also in the coding of the site itself.

This requires making use of **meta tags**, which help inform the search engines of what your site is all about. Of course, appropriate use of meta tags also gives your SEO a big overall boost. Search engines are better able to index these pages, providing a big increase in visibility and rankings.

Meta tags are HTML tags you place on the coding of the page. They can include **title tags**, **description meta tags**, **heading tags** and **image tags**, among others.

Certain types of site software provide toolkits that make it easier to add these tags to your site without messing around with HTML — the most basic website programming language.

Title tags: These are extremely important indicators that inform search engines what your page is about. Things to keep in mind:

The title tag is not to be confused with your *page heading*. It's part of your meta data and the title users see at the very top of your web browser. It's also the title people will see for your site on a SERP.

Google typically only displays the first 50-60 characters in a title tag, so don't go over 60 (including spaces). Because you have a limited number of characters, your title is precious real estate. Think like a user and type in search terms you know the user will use.

Some users will only look at titles, so your title needs to be something that will compel the user to click through when your result shows up in SERPs.

Titles for each page also must be unique; they cannot be repeated anywhere else on your site. The best SEO practice is to include your keyword in your page title, preferably at the beginning of the title.

For your field, as with all tags on your site, something specific to hospice and related palliative care obviously should be incorporated.

Meta description tags: These describe what your page is about and are displayed along with your title in search results. Obviously, how your site is described will be a huge influence on whether a searcher clicks through to your site.

Google will display 120-158 characters (including spaces) for a meta description, so again, don't make it too long. Make it something that's direct and compelling, and include words that users are likely to use to find the page.

Get your primary keywords in there, but avoid keyword-stuffing, and also try to include a call to action such as *read more about…*, *find out...*, or *learn all about….*

An enticing description will increase your click-through rate. It also lets you stand out from other sites that show up in SERPs.

Heading (H1) tags: These tags are intended to indicate the most prominent (and presumably most important) text on a page. These are followed by other H tags, H2 through H6, ordered from most important to least. These are important from an SEO perspective, but they also make the page easy to read.

Think about a magazine. Important things, such as the magazine's title or a front-page feature, would be in H1. A smaller article's title might be in H2 and a subtitle in H3.

Google expects a webpage to be laid out with a series of headers that makes sense to the reader. These tell Google and other search engines what is most important.

It's important not to go overboard on H1 tags, just like you don't want to go overboard on things such as bold or italics.

Emphasis is used to *differentiate* things as being more important than other things; ***if everything is "important," nothing is***.

Abusing H1 tags will definitely get a webpage penalized by Google and other search engines, so creating multiple H1 tags on a page purely for the purpose of trying to improve your search engine ranking is definitely frowned upon. It's generally best to stick with just one H1 tag per page.

Image tags: While both human visitors and search engine spiders can interpret words on a page, images themselves are (for the most part) only able to be interpreted by actual humans.

To let the spiders know what's being seen in an image, you need to add image tags. These are also known as **alt-image tags** because the text you include in them will be shown to your human visitors if the image fails to load for some reason.

Search engines crawl these tags as well, so by including hospice-related keywords (when natural and appropriate) in these tags, you're once again improving your SEO for people who search for these keywords.

As usual, don't stuff keywords into tags willy-nilly, because that can blow up in your face.

Section 3:

Reviews and Reputation

CHAPTER 5
Review Sites and Reputation Management

The internet lets you tell everyone in the world how awesome your services are. You can go on about your dedication to patient services on your website. You can develop advertisements that boast about your being the very best hospice in your neighborhood or city.

And if that spiel is legitimate, worded well, and backed up by strong testimonials, people will most likely accept that your assessment of your stellar reputation is genuine.

However, there's another side to dealing with reputation issues in the age of the internet:

You're not the only one who gets to weigh in on your business.

Whether you like it or not, members of the general public will have their say on their experiences with your hospice.

Of course, you understand the importance of CAHPS scores, and this chapter isn't intended to suggest online reviews from broader platforms make CAHPS less significant.

But it's also true that most of the general public doesn't know what a CAHPS score is. Virtually everyone knows what a Google review is — and Google makes a point of ensuring it's overtly displayed to families seeking your services.

As I noted earlier, customer/client review sites exist to aggregate information from people who have tried out a business and wish to write a review.

While some people enjoy writing reviews of almost every establishment or service provider they interact with, most only take the time to write a review if they had a spectacular experience — or a truly horrible one. For this reason, all businesses need to be ready to do a great job for every customer or client.

That's not a bad idea in theory: With its online reputation potentially at stake every day, a business presumably will make a point of doing a great job with every interaction. That's extra impetus to go the extra mile, and it's better for clients as well.

Of course, like anything else on the internet, reviews can be gamed.

As terrible as it sounds, certain less scrupulous business owners try to artificially inflate their positive reviews by registering on these sites under a variety of different names.

Even worse, these business owners often *pan* their local competitors — whether or not the criticisms are deserved — under other names. (Thankfully, this doesn't happen as often among hospices as in other industries, but it certainly happens occasionally.)

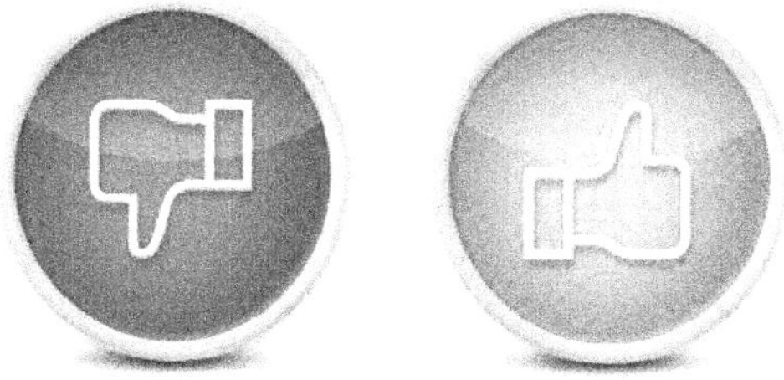

Also, unfair reviews are sometimes posted by people who don't necessarily have a stake in the businesses. Some people just get a kick out of ripping companies — again, regardless of whether such criticism is actually warranted.

Others try to trade super-positive reviews for free stuff, discounts, upgrades, and other perks. All these underhanded strategies can influence the legitimacy of online reviews.

Because of this, most of the major review sites, along with Google, set up algorithms that are intended to seek out and hide or remove reviews that are false or illegitimate. These algorithms have had mixed results, but they've at least applied some controls to a process that can be fraught with misbehavior.

The major sites also include some options to allow businesses to contest unfair or inappropriate reviews, and all let the business comment on reviews that appear, providing the company an opportunity to place the criticism in context and (if desired) to communicate a desire to address any legitimate concerns that were noted.

I'll talk about some of those options in a bit, but what's most important to take from this is that *you must be vigilant in reviewing your hospice's online reputation.*

This means paying attention to whatever people are saying about your hospice on review sites, in blogs, in news reports, and anywhere else.

People who contract for something such as palliative services are somewhat more likely to post about the experience — whether positive or negative — than someone who simply bought a bagel from the coffee shop down the street. (Granted, there are people who do this, and we wish them all the best.)

If a reviewer hits you with a particularly negative review, it's *critical* that you not let it sit out there without any comment or response. If several bad reviews go unanswered, that can be *extraordinarily damaging* to your reputation. No one wants to provide a referral or contract for services with a hospice that's been shredded in online reviews.

Remember at the start of this book when I explained how more people every day are getting information about local service providers from their smartphones?

These devices make it unbelievably easy to quickly check out reviews when a potential client pulls up your Google listing or a Yelp app. And remember, Bing also integrates Yelp reviews into its searches.

These days, when someone simply clicks on your business in a maps application or does a search for nearby businesses and comes across your hospice, those reviews are sitting right there to be read.

No matter how well you promote your firm, if several negative reviews pop up any time someone check you out, you're in big trouble.

You absolutely *can't afford to ignore online client reviews*. Good ones can be incredibly helpful, but bad ones can be devastating.

You *must* proactively generate positive online reviews. They're critical for success.

I realize the process requires more nuance in hospice, but it can be done. Many of my clients have been very successful at it—it's simply a matter of dedicating yourself to the process and initiating it.

But before we go any further, we need to talk about:

Review Gating

Yes, it's a strange term. Don't worry, I'm about to explain it.

Review gating refers to the process of using specialized software to filter out potential negative reviews and only let people publish positive reviews of your firm.

The typical process works like this: You send an email to an individual (for hospice, typically a family representative of the patient) asking whether they were satisfied with the services you provided and whether they'd like to publish a review.

If they were completely satisfied, they're guided to a clickable link where they can easily write a quick review.

If they were *not* entirely satisfied, they're guided to clickable link that lets them contact you directly about their concerns — bypassing the publishing process.

Obviously, this let businesses avoid negative reviews by "heading them off at the pass." Which is why Google banned the practice back in 2018.

You don't want to be involved in this process, because as I noted before, the last thing you want to do is upset Google. If you get caught, you'll be penalized.

With that said, there *are* software options that allow people to publish negative reviews if they wish, but they *also* provide the option to reach out to the business first.

By doing this, a dissatisfied client has the opportunity to address concerns before reviewing, if so desired, but is not actually barred from leaving a negative review.

We've found that many families are fine with contacting the business directly before leaving a negative review, and after working things out, they either choose not to leave a negative review — or actually leave a positive one! At HHM, we have proprietary software that facilitates this process.

If a marketing agency claims it can make negative reviews "disappear" or entirely block them from occurring, don't work with that agency. No one can legitimately promise that.

However, there are means by which you can negotiate with people who have left negative reviews and often get them to remove the review or even choose to modify it to be more positive. Also, all review platforms have processes for removing truly fake reviews.

While there are lots of review sites online — some of which exist only to cover particular types of businesses or niches — it's best to focus on Google and Yelp (very much in that order). Let's knock out Yelp first:

Yelp

Google reviews have severely cut into Yelp's early dominance, and as I noted earlier, Yelp doesn't play a very large role in fields such as hospice. However, it remains important to keep an eye on any Yelp reviews and ensure they're positive.

As one of the first sites to marry local search, user reviews, and social networking aspects, Yelp has grown astronomically since its founding in 2004, and its reach has expanded with ridiculous speed in recent years.

Yelp also has a reputation system that lets visitors see which contributing users are the most popular, respected, and prolific. This can help visitors judge how legitimate reviews from these users tend to be, among other things.

Business owners can also communicate with contributors who post reviews on their page through messages or public comments.

Having reviews of your hospice on Yelp — especially positive reviews — does help, at least a bit.

As I mentioned earlier, the idea of a review aggregation site compiling plaudits and criticisms of various businesses is a controversial one.

Some business owners have questioned Yelp's own credibility, especially after allegations emerged that some Yelp salespeople might have offered to hide negative reviews of businesses that paid for advertising sponsorship contracts. Yelp has refuted these allegations, but they're pervasive enough to be concerning.

Certain businesses also question the effectiveness of Yelp's review filter, which is intended to prevent and remove illegitimate reviews. In certain cases, the businesses claim, reviews that would seem to be clearly false show up next to "real" reviews. In others, honest-to-goodness legit reviews get snagged by the filter and are not displayed.

In the same way that Google won't disclose the specifics of its search ranking algorithms to prevent them from being gamed, Yelp won't disclose the specifics of its review filter algorithms.

Several class-action lawsuits have been filed against Yelp by businesses that have had positive reviews removed but have been unable to get negative reviews similarly redressed.

One frustrating factor with Yelp is that it no longer allows businesses to directly solicit reviews. It wants satisfied clients to decide entirely on their own to review a business.

I don't recommend trying to work around this regulation, because if Yelp catches you soliciting reviews, it can penalize your hospice on its platform.

With that said, there's nothing wrong with asking families for reviews in general, and if *they* ask whether to do so Yelp, there's nothing prohibiting you from saying "that's fine."

Advertising with Yelp: While the company continues to contend that advertising with Yelp won't affect the display of organically created user reviews (either positive or negative), it seems safe to assume that advertising on Yelp can't hurt.

At the very least, the company makes no bones about the fact that your Yelp page will be better promoted and get more visibility if you're an advertiser.

However, that increased visibility might not be worth the return on investment for most small businesses. Many marketing experts believe Yelp's paid ads currently cost far more than what you receive in return.

The numbers can get a little confusing if you're not familiar with **cost per impressions**, which is the model Yelp uses for the ads. Much like traditional advertising, an *impression* merely means that someone has *seen* your ad. You're paying for the ad regardless of whether someone clicks through to your site.

Cost per click, by comparison, only costs you money if someone clicks through.

With Yelp ads, the cost per impression (CPM) is exponentially more expensive than it is with standard CPM advertising. It's also much harder to track how effective your advertising is when you can't get click-through numbers.

Yelp is also requiring a one-year commitment for its most favorable rates, which aren't particularly low in the first place.

In general, I don't recommend Yelp advertising at this time for most small businesses, including hospices. If you have plenty of extra marketing money to burn and want to give it a shot, it's an option to consider. But I think you'll get better ROI from other initiatives.

Google reviews / Google Business Profile

I covered GBP extensively earlier, so I won't get into it deeply here.

The thing to know is that people can review your hospice through Google, and these reviews are incredibly important. With the vast majority of people using Google to research local businesses, these reviews are among the most prominently displayed to users.

The quality *and* quantity of Google reviews your hospice receives also plays an indirect role in SEO. The short version is that Google's spiders can tell which businesses are reviewed a lot, so they grant those websites more authority. And more authority means higher search rankings.

Note that if you haven't verified your business with GBP yet, your information won't necessarily show up on Maps, Search, and other Google services. Additionally, *only verified businesses can respond to reviews*.

So if you're not sure whether you've gone through verification, stop everything (seriously, just put this book down or close your e-reader) and do it right now.

Google systems are integrated by default into all Android-based devices, so you want to have the best reviews possible on your Google listing.

When someone Googles your business, your verified GBP listing typically will appear on the right side of the SERP. One of the first things people will see is your Google review rating: your overall score (on a five-star scale) and how many reviews you've received.

While it's great to have a handful of five-star reviews, surveys have shown that it's actually better to have a *higher quantity* of reviews, even if they're not all five stars!

In other words, people generally have greater trust in a business that has (for example) 25 reviews with a 4.7 overall rating than one that has seven reviews with a perfect five-star rating.

You always want five-star reviews, of course, but don't let "perfect be the enemy of the good," as they say. Work hard to generate *a lot* of Google reviews.

Like Yelp, Google Business Profile uses a reviewer algorithm to ensure the display of the most legitimate reviews.

Google provides an option to flag reviews as inappropriate if they violate its guidelines, but as with Yelp, it can be hard to simply have a negative review removed if it falls within the guidelines. You can respond to a negative review with comments to provide your side of the issue.

Garnering more reviews

While negative reviews obviously are bad for your hospice, the best way to counter them typically is to solicit reviews from satisfied families. They're obviously far more likely to rate your business positively, and as I noted before, the quantity of reviews your hospice receives will provide a huge boost to your rankings in local search results.

Like it or not, there's no way to get rid of online reviews. They're simply a reality of the times. So instead of fighting them, do your best to get the most positive reviews you can. Ways to do this include:

Solicit reviews on your website: I mentioned before that Yelp prohibits businesses from directly soliciting reviews, so don't do that. However, there's nothing wrong with asking satisfied families to review you on other platforms.

You can create a page on your site specifically intended for reviews, and that can have a simple form they can fill out. The easier you make it for the client, the more likely you are to get a review — particularly a positive one.

Put it in an email: Again, including a simple link to your own review form or that on a popular platform makes it easy for the client. Just be sure you're being careful not to spam your client with emails, which could very easily prompt *negative* reviews.

Monitoring and responding to positive and negative reviews

It's never fun having your services criticized, regardless of whether the feedback is fair. When it comes to an online review, the impact of a complaining client is far heavier than that of a card in a client feedback box. It can imperil your hospice now and in the future.

This is why it's so critical to stay on top of how your business is being reviewed online. Remember that *reviews are good* for your hospice in many ways, so you shouldn't just put your head in the sand or try to stay under the radar. You do a good job, and your clients know this. The key is to get them to make this known, as I discussed in the previous section.

One thing you absolutely must do is *monitor all reviews you receive.* It's not as hard as it sounds. In fact, there are services available that can help monitor all the major review sites and let you review them together, and your internet marketing professional can help with this. If you don't know what's being said about you online, you're powerless to adapt to it.

While it's most important to respond (and respond *quickly*) to negative reviews, the importance of responding to positive reviews is often overlooked. When people take the time to say something nice about your firm, they like to know that you've seen the review and appreciate it. Thanking them and promising to do a great job will go a long way toward attracting new clientele.

Also, if the only reviews you respond to are the negative ones, that leaves a poor impression on everyone who sees that page. Thanking people for glowing reviews and responding positively to critical reviews will always paint you in the very best light.

Granted, it can be challenging at first to find a positive tone in replying to a negative review, but it's not too hard once you get used to it.

After a cooling-off period, respond privately to the reviewer: You don't always have to start off with a public comment. On Yelp, for example, you can reach out to the concerned client first through a private message. Perhaps the criticism resulted purely from a misunderstanding or miscommunication that can be worked out.

Whatever the case, give yourself a little time to let your emotions ebb. It's natural to be upset by a negative review, but resist the temptation to overreact.

Instead, take a deep breath and evaluate the client's complaint. Determine what has been (or can still be) done to address the issue. Engage the client with a short and positive private response — either through the online review platform or via email — that recognizes the criticism and discusses ways to remedy the situation.

By taking the issue offline, you can prevent a potentially ugly back-and-forth discussion that could draw even more unwanted attention to the review. If successful in resolving the issue offline, you can politely ask the client to revise the review to note that the complaint was addressed or even delete it altogether.

If that doesn't help, provide a public reply: Even if you work things out privately, the reviewer might not elect to revise or remove the original negative review. Don't badger the reviewer; this might only lead the reviewer to further criticize your hospice on a follow-up comment or another review.

If you believe the issue was resolved, note that you reached out and successfully addressed the client's concern.

If you believe there were legitimate inaccuracies with claims made in the review, offer a short response that corrects the facts. Again, keep it positive and professional. Maintaining the high road and appearing conciliatory to client concerns is more important than establishing "your side."

Another thing to consider is that you should do everything in your power to discourage negative reviews in the first place (while encouraging positive ones). The foundation of this is simple: Do what you do incredibly well.

However, no matter how well your hospice is run, there still might be clients who are unhappy. That's just the nature of business and the broad spectrum by which people interpret their experience. Some people are easy to please; others never seem to be satisfied. Two people could have the exact same experience and interpret it two completely different ways.

Many times, a client will note concerns directly with you before choosing to post them publicly. It's critical to seize this opportunity while you still have a chance to allay these concerns. Respond quickly and work out a resolution before a negative review is posted.

Over time, negative reviews can cost you much more than what would be involved in proactively remedying a client's complaint.

CHAPTER 6
Email Marketing

Think about the most tech-wary person you know, a true technophobe through and through.

Can you think of anyone? Someone who has truly let the digital revolution pass him by. Doesn't have a smartphone, much less the two (or even three) some people carry today. Doesn't have a laptop, much less a tablet. Has no interest in X (Twitter), doesn't care about Facebook, wouldn't know Yelp from yogurt.

At the very least, your friend almost certainly has a home PC and uses email.

In reality, almost everyone in America uses email and checks it regularly. Even people who rarely leave home check their email on a constant basis.

Most modern email systems provide pop-ups and alerts to let you know you have a new email. Preschoolers have it. Great-grandmothers have it. With postal rates continuing to rise all the time, people continue to shift from communicating by snail mail to email.

And even if you don't own or use a smartphone, it's a fair bet your mobile phone — even one manufactured a decade or more ago — has the capability to receive and send email. Your email provider is most likely free to use, making it all the more valuable. Email access is easy, it's free, it's omnipresent.

Of course, email also produces a lot of spam, which is another thing entirely, but I'll get back to that.

Email marketing can be tricky in some ways, but it's a marketing initiative no business can afford to ignore, and (you already see this coming) that certainly includes hospices. You have a free way to communicate with prospective clientele who opt-in to your emails.

Can emails sometimes get lost in the morass of communications people get in their inboxes every day? Sure, but no less than direct mail advertising gets lost in the stack of junk that piles up in the mailbox. And direct mail costs *money*.

Email is by far the most cost-effective way to deliver marketing messages for your hospice. You can send personalized, targeted, and interest-specific messages to a large number of people.

But there are other advantages. You can include links in email that prompt the recipient to click through directly to your site/landing page. Also, studies show that more than 80 percent of the email you send is opened in the first 48 hours after delivery, so you can get a fairly immediate response or action based on it.

First, avoid spam

Before getting into some email strategies, let me repeat something I touched on earlier:

Never spam.

Never, *ever* spam.

If you're sending out a mass email you think *might* be spam, it's almost certainly spam. Don't send it.

Sending out unsolicited emails is the very best way to get your business identified as a spammer, ensuring your emails a permanent grave in the ever-more-sensitive spam filters developed by Gmail, Outlook, etc.

When was the last time you even *checked* your spam filter, much less actually clicked on something you didn't recognize in it?

Spam emails are typically automatically deleted every 30 days or so for good reason: They're not beneficial, and no one wants them.

Even if your unsolicited email sneaks around the filter and reaches a recipient's inbox, most email providers provide a one-click method to report an email as spam. If your email looks spammy, you can expect this to happen.

Oh, and one other thing: *spamming is technically illegal*. So there's that.

Sticking with professional email standards keeps your emails legally compliant and ensures great relationships with the people who receive your emails.

An internet marketing professional can help you understand all the ways your email could be classified as spam, though they mostly fall under common sense. Even if you always send emails to people who have opted in, there are two critical things to consider:

To send commercial emails that don't violate the law, every email should include two things: the *physical address* associated with your business (a P.O. box is accepted as long as it's associated with your business's actual location), and an *unsubscribe option*.

The unsubscribe process also must be accomplishable by replying to a single email or by visiting a single web page.

You're also required by law to remove anyone who unsubscribes from your email list permanently within 10 days of the unsubscribe request. You can't add that person back without the recipient's explicit permission.

Getting permission

If you haven't received permission from a recipient, your email almost certainly will be considered spam. So always get permission.

Technically, there are two types of permission: **implicit** or **explicit**.

Implicit permission occurs when someone shares his or her email address with you for some reason, such as filling out a contact form on your website for more information regarding a particular service or item.

That's opened the door for you to contact the person for that reason, but you haven't actually asked permission to continue to contact the person down the line with product offers and the like (unless that's clearly noted when the person signs on).

It's much safer to get explicit permission: Tell the person what to expect when by providing an email. If you plan to send promotions and other communications, be straight about it.

Once you've compiled your list (more on that in a moment), don't share it, don't sell it, don't trade it. Those are surefire ways to destroy your reputation with your clients.

Do I really have to be that careful? In a word, yes. As few as two spam complaints per 1,000 clients can prompt internet service providers to block you from sending emails to their customers.

Even when you've received explicit permission, your emails can be interpreted as spam depending on how they're constructed or how often you send them.

In a sense, it doesn't matter whether the recipient asked you to email in the first place. If the emails appear too frequently, don't appear legitimate, or don't provide useful information, they can soon be interpreted to be spam.

Putting your list together

It's harder these days to compile a large list than it used to be, which is understandable: People are more careful about giving out their email address for fear of getting crushed by communications every day. Ever overlook an important email because it was buried in a bunch of inbox chaff? It's not a fun feeling.

In reality, you don't need a huge list for your email marketing to be effective.

The best lists contain the names of referral sources who respect others' privacy and interested prospects who know you and your hospice well enough to recognize (and want) your communications. Ultimately, *quality* is much more important than *quantity*.

Unless your list is going to be extremely short, it's best to use some sort of database (or software specifically designed to organize emails and contacts).

The database for your email list can be part of a client database you currently use, or it can be something entirely separate. That said, it's *critical* not to get confused about which clients (and prospective clients) have agreed to receive emails from you.

A contact form on your website is one of the most common ways to compile an email list. If you're having trouble getting visitors to fill out the form, some incentives can help: *Join our email list and receive 20 percent off your first purchase.*

Design a great email

An effective email starts with the header. This includes the "From" line and address and the subject line.

The "From" line doesn't simply have to be your email address per se. You can set this up to read as your full name, or your firm's name, or both.

If another name (such as your hospice website's domain) might be more recognizable to people who have signed up for your emails, it's best to use this name. You don't want recipients to be confused about who's sending them emails.

The **subject line** is incredibly important, because this is the most information that will be provided before someone opens your email. It must compel the recipient to want to know more.

Subject line real estate is valuable, because most email providers only display the first 30 to 50 characters, and most mobile devices only show the first 20 to 30 characters. You need to get your point across concisely.

Normally this is where we'd suggest a strong sales pitch, but that might backfire in a more sensitive business such as hospice services.

As with any professional communication, it's critical to spell words in your subject lines correctly and avoid spammy elements such as excessive punctuations (especially exclamation points), all-caps, capitalizing the first letter of every word, or using "Re:" (when it's not actually in response to anything).

Also, ambiguous or deceptive subject lines such as "Hi!" or "I've missed you" are always to be avoided (and would be inappropriate for hospice marketing regardless).

Be sure to brand your emails with a consistent template or a consistent color scheme. Including your hospice's logo also is a must. It should look like what it is: a professional communication from a professional provider of hospice services.

Visual elements: Marketing pros will tell you there are a lot of psychological elements you can use in your design to draw the eye where you want it to be. That's absolutely true, but it's beyond the scope of this book.

You should know the basics: Be sure the email is clear and easy to read. Be sure the most important aspects (particularly the call to action) are eye-catching and direct. Add visual anchors such as images (within reason) to break up the copy and make it appealing to read.

(Note on images: Without getting too technical, it's much better for your images to be referenced to a remote folder than embedded. Emails with embedded links are much more likely to be filtered as junk. A marketing expert can help you understand the difference.)

Content of your email

Be personable: As with your landing page copy, you don't want to sound too academic. Be professional and caring, but don't be stuffy.

Links: If you're not linking to your landing page or a page that's specific to your offer in your email, you're wasting a great opportunity. Even if your email is purely communicative, there's no reason not to provide a one-click opportunity to get more information on you and your business in the body of the email.

An informational email works well, too: Maybe you're not looking to promote a particular service at this time. That's fine. Is there some information about hospice services that would be (truly) interesting to your potential clientele? If so, that's enough reason for an email.

Don't waste people's time by sending these out constantly or for minor matters, but delivering useful information is a great way to stay in people's minds, along with a great excuse to include links to your website.

If you don't have the time or resources to create your own emails, there are lots of great copywriting services that can help you achieve your goals, and many are very reasonable.

Monitoring your email results

Again, I won't get into all the technical aspects here, but there are great services that will allow you to easily track your return on investment from business emails. They let you review lots of useful data, including:

- Which emails bounced
- Why they bounced
- Who opened the emails
- What links the recipient clicked
- Who unsubscribed
- Who forwarded your emails

This data can go a long way toward helping you refine your email marketing efforts and improving conversions.

There are more elements to successful email campaigns, but understanding these basics will help you determine the best way to use email to help your firm succeed.

Email lists: Should you purchase them? (Spoiler alert: no.)

A precompiled email list sure sounds like a good deal when you're marketing your hospice firm. You have a list of emails for people who are prequalified to be interested in what you do. What could go wrong?

Um… just *everything*.

The first thing to remember is that even if you worked with an email list provider for a list that has been shaped to include only clients who meet certain demographic or psychographic (personality, interests, lifestyles, etc.) standards, *it's not really an opt-in list.*

No one opted in for communications regarding your services. The people on that email list simply opted in to an email communication from someone at some point in time (like the list provider).

When you're emailing people who haven't shown an interest in your firm, that's something most people consider spamming — and you're not likely to see much return on those communications anyway.

Also, there's really no such thing as a good email list that's for sale. Typically, any addresses that once had value already have been spammed by a variety of other businesses.

Yes, there's that word again: **spam**. We're printing it in bold, even though we rarely do that (extra tip: use emphasis tools such as bold sparingly so they don't lose all meaning), because it's so important to understand that contacting people through a purchased email list almost always equates to spam.

If your firm is defined as a spammer, your email deliverability and the reputation of your IP address almost certainly will be harmed. Organizations dedicated to combating email spam have a tool called a **honeypot**. It's a planted email address that, when emailed, identifies the sender as a spammer.

There are other spam traps that identify if an email address is old (or no longer valid) but still receives consistent traffic.

There are effective ways to build email lists through proper means. For the sake of your firm and its reputation, don't go the purchased email route. The rewards are few and the risks far too great.

Section 4: Monitoring and Measuring

CHAPTER 7
What Gets Measured Gets Done

Yes, "What gets measured gets done." It's a well-known saying, and properly so. *It works.* When you're constantly monitoring and measuring your results, you can determine what's working and what's not. You can find the places where you need to revise, tweak, or even go in an entirely new direction.

Here's the beautiful thing about today's digital world: Monitoring and measuring your marketing efforts is now incredibly fast, unbelievably precise, and unexpectedly easy.

Not only that, but you have far more flexibility now in what you do with the information you get from your analytics. You'll know what you need to do and be able to do exactly that on the fly.

You can make changes to your website that go live immediately. (Yes, it can take some time for spiders to crawl the site, but changes you make will be seen by human visitors right away.)

You can revise a PPC ad so that leads see the new information the very same day. When your analytics show that your clients are excited about a particular promotion, you can showcase it on social media.

Google Analytics

This is the best known and most popular of the user-friendly measuring tools, owing at least in part to the fact that so many businesses use Google advertising and other services.

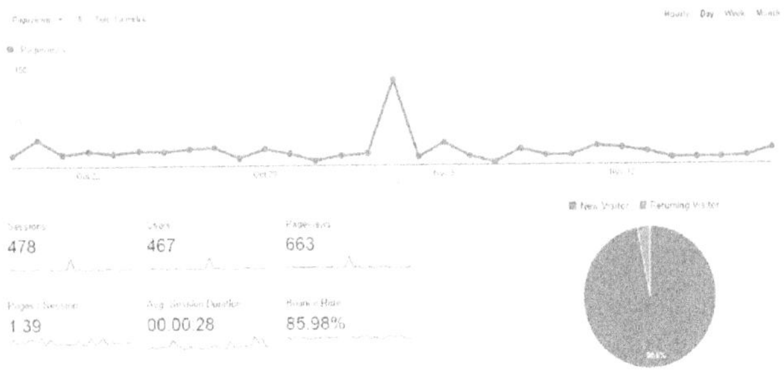

In essence, Google Analytics generates detailed statistics about a website's traffic and traffic sources, along with measuring conversions and sales. This makes it an incredibly useful tool for both marketers and business owners. The basic service is free, but there's a premium option available for a fee.

The service can track visitors from a broad variety of sources, including search engines, social networks, direct visits, and referring sites. It also displays advertising, pay-per-click networks, email marketing, and digital collateral such as links within PDF documents.

Google Analytics is completely integrated with Google Ads, making it easy to review online campaigns by tracking landing page quality and conversions.

Target goals can include sales, lead generation, viewing a specific page, or downloading a particular file, among other options.

Some of the many things you can analyze

Whether you use Google Analytics or another site-monitoring tool — the former is practically a requirement, and other tools are best used in combination with Google Analytics — here are some of the many things you can monitor:

- **Number of visits**
- **Number of people who actually visited your site:** A person may visit your site more than once, and each time, that's a separate visit.
- **Number of pages per visit:** Each time a person visits your site, the visitor might view a single page or go on to view dozens of them. (If you've populated your site with great content, hopefully the visitor is reviewing it in detail.) The greater the number of pages viewed per visit, the more people are actually exploring the information it contains.
- **List of most visited pages:** This indicates the number of visits received by each individual page. You can then use this list to analyze which themes, products, and information grab your clients' attention the most. It also includes your campaign landing pages, providing you with final statistics for each of your marketing initiatives.

- **Length of stay:** This indicates how long on average your visitors stayed on your site, thus indicating how well they relate to your content. The longer your visitors stay, the greater their interest. This gets to the issue of **bounce rate**, which I'll touch on in a moment.

- **Origin of visitors:** This identifies what site someone was viewing on the internet immediately before visiting your site. Think of it like a flight: One city's airport is the place of departure; the other is the place of arrival. It's beneficial to know where your visitor arrived from — that makes it much easier to know what prompted them to check out your firm.

- These origin statistics can show **direct visits**, in which visitors enter your URL directly into the browser; **site referrals**, in which a visitor comes to your site after clicking on a link at another website, a **blog** or **social media network**; and **search engines results pages**.

Tracking phone calls

Businesses that mostly do e-commerce and sell products and services online aren't so concerned with phone calls. They do the vast majority of their business digitally.

In fact, many products are sold online without any direct involvement by the person selling them. An automated purchasing system is set up, the customer purchases the item, and it's automatically delivered digitally (if not a physical item) or goes into a queue to be delivered from a drop-shipper or wholesaler.

For a hospice, of course, it's an entirely different story. While you might (and should) get inquiries by email or through a form on your website, if you do receive a direct inquiry from a family, it's more likely to come in over the phone.

Sure, some hair salons have online registration systems, just like restaurants have online reservations (either on their own site or through an app such as Open Table) and many other service providers.

And there's no question that today, most millennials and (particularly) Generation Z are *far more comfortable* handling their business digitally as they are over the phone.

But when looking at all demographics, even today most people making an appointment with a local service provider do so over the phone.

That's particularly true in an industry such as hospice, where such matters obviously are of a very personal nature. Making arrangements for a family member needing hospice care is the sort of thing one does in person or over the phone. (Of course, there's also Zoom, Skype, and other types of real-time chat, but I'll stick with the basics here.) This is why understanding how to track phone calls that originate from your website is so important.

Many people now make phone calls simply by clicking on a link. When they do it on a desktop computer, typically they have it set up to call through Zoom, Skype or some other VoIP (**voice over Internet Protocol**) service.

These days, however, many more people will be visiting your website — or your Google Business Profile page, Yelp listing, Facebook Business page, a PPC ad, etc. — on a digital device that's *also* a phone.

Because your phone number is also a link that allows people to call you with a click, they can contact you *immediately* directly from your listing. The easier you make it for the lead to contact you, the better.

That's why having a *clickable* phone number (just like a clickable email address) everywhere you have a Web presence is so important.

When your phone number is clickable, an experienced internet marketing pro can easily track phone calls made from your site and listings. You can see how many leads are coming directly from your site in the form of phone calls (emails also are easily tracked, as you might assume).

You can see what times of the day the site is most effective in getting phone leads, how long people visited the site before calling, what page they were viewing before they called, and so on. This lets you optimize your site to convert the most traffic into leads.

Bounce rate and site engagement

As I indicated earlier, today your website can't merely exist because "everybody has one." It's not just there to display some elegant pictures of your hospice, along with some basic contact information.

Your website needs to be *actively engaging* its visitors. It must be designed to communicate trustworthiness, engage the visitor, sell the visitor that you're the best solution to a particular problem, and convert the visitor into a client.

You can implement all the best practices toward creating an engaging, high-converting site, but only through analytics will you know how well it's working — and which aspects of the site are delivering the conversions (and which are not).

The first thing most experts focus on when assessing site engagement is the **bounce rate**. Though the definition for this can vary a bit, I'll stick with the Google Analytics definition: It's the percentage of visits in which the visitor only views *one page* before leaving the site.

For certain businesses, having a high bounce rate isn't a terrible thing; they might be trying to communicate something very simple with a very basic single landing page. They might not even have more than one page on the site.

Bounce rate can be broken down in many ways, including factoring which visitor origins have the highest bounce rate and which have the lowest.

The other concern regarding bounce rates relates to SEO: Google and other search engines take your bounce rate into account when ranking you. It's another key reason to design your site in a way that invites the visitor to stick around and check out additional content.

Reducing bounce rate

If you've set up your site in a legitimate manner with white-hat SEO, and at least spread some solid content across several pages, you don't have to worry about the worst bounce rate percentages.

Those terrible numbers typically are reserved for sites that attract visitors through misleading (black-hat) means.

They persuade people to click through to the site by implying there is a diamond in the stocking when they click through, but what the visitor finds is a lump of coal. The visitor immediately clicks out of the site, which is why bounce rate is such a major factor in search engine rankings.

Still, the better the bounce rate, the better for your business. Ways to improve your bounce rate include:

Browser/mobile compatibility: If your site isn't optimized for mobile devices, you'll get *a lot* of bounces from visitors trying to access it on phones. Also, be sure your site displays properly on all the major browsers — especially Google Chrome, the industry leader.

Pay off your ads on your site: When someone reads an ad that promises one thing, but clicks through to discover another, that's a quick bounce. Be sure your site is set up logically and intuitively, and be sure your site delivers exactly what your ads promise.

Don't annoy the visitor: If you hit the visitor with pop-up ads, chat windows, a blatant call to action, and more right on the landing page, kiss that visitor goodbye. Obviously, a hospice website with a tacky interface is going to put off potential referrers and clientele in no time flat.

Keep your landing page attractive and welcoming. Let the visitor feel "at home" and interested in exploring before placing demands. It never hurts to think of a virtual visitor as a guest in your home.

Check your page load times: If your site uses extremely complicated design or lots of graphics, some pages could take a while to load — this is a known factor in high bounce rates.

You can test page load times through analytics. If your pages are taking substantially longer than average, it's time to do some redesign.

There are ways to streamline the mobile version of your site so that large files, for example, do not automatically display.

The visitor can be provided an option to display these files if desired, which might be the case, for example, if the visitor is viewing the site on a tablet or large-display smartphone over Wi-Fi.

Improve your content: This is a general tip, but it's one of the most important. Small things such as spelling and grammatical errors can cause a visitor to lose faith in your business and bounce out. So can something that could be considered polarizing, such as political statement. High-quality content returns low bounce rates. Low-quality content returns high bounce rates.

The five basic traffic metrics

When analyzing your website's traffic, there are five basic metrics to understand:

Sessions: A session is another word for a single visit. It's one person visiting your site any one time. If someone visits your site 15 times, those visits count as 15 sessions, whether she stayed on the site for 5 minutes or 50.

Unique visitors: A unique visitor is one person visiting your site any number of times during a defined period. If someone visits your site eight times in a week, that person still counts as one unique visitor.

Page views: A single page view is any one visitor viewing one page of your site, one time. The page must have a unique address (URL). If someone visits your landing page and then clicks a link to your About Us page, those are two page views.

Time on site: This is the total amount of time one visitor spends on your site in the course of a single session. Average time on site is a critical measure of visit quality and visitor interest.

Referrers: If someone clicks a search result link to reach your site, the point of origin — the departure airport I talked about earlier — is known as the referrer.

There are many ways in which a marketing professional can use these metrics to help improve your site's traffic and conversions, but the details go somewhat beyond the scope of this book.

What's critical to understand regarding monitoring and measuring is the digital age lets you precisely understand every aspect of your site's effectiveness and your marketing campaign — and it lets you make virtually instantaneous changes to optimize your efforts.

Knowing which referrers tend to send the most traffic your way can help you decide where to focus your marketing and promotional efforts.

Of course, it's just as important to know how well visitors are converting who arrive from particular referrers — another statistic that can be easily mined from the data.

Analytics essentially provide you honest feedback on all your efforts. People might tell you they like one thing or another, but the proof is in how long they spend on a page, what they choose to click on, whether they decide to move forward and convert to a client. The digital age lets you see all that information in a way that can power your success.

Section 5:

Working with a Marketing Professional

CHAPTER 8
Finding a Qualified, Principled Marketing Professional

Frankly, many marketers remain behind the 8-ball when it comes to understanding how to promote their firms online.

You'll want to find a company that specializes in online marketing, but it's just as important to be sure the company uses appropriate methods that will keep you on the good side of Google, other search engines, and laws regarding spam and legitimate marketing methods.

But this consideration is most important of all: You want to work with **full-stack marketing agency**.

What does that mean?

I'm glad you (hopefully) asked.

The short version is that you want a marketing agency that can do it all, that has a wide spectrum of "weapons" in its arsenal and can employ them as needed.

One approach most definitely does not fit all when marketing hospice. Different firms have different needs, different budgets, different goals.

They have different levels of competition in their particular areas. Speaking of areas, the geography of a particular hospice plays a big role in which marketing methods will work best.

Also, many so-called marketing agencies don't do remarketing. Or hyper-geotargeting. Or review and reputation optimization. Or website conversions optimization.

A (frankly shocking) number of agencies are happy to take your money and add your hospice to their "one-size-fits-all solution." And honestly, it doesn't work.

It's also true that you don't want to waste money on marketing services you don't need — or don't fit your particular goals at the time.

A reputable full-stack agency has all the options available, but it *also* won't try to pad its wallet by requiring you to accept a "full suite of services" if some make zero sense for your particular situation.

Instead, a legitimate agency will take the time to understand your exact situation and work *with* you to create a campaign that includes the services with the best ROI and most effectiveness for you.

When scrutinizing a marketing agency, first make sure it effectively employ all the tools listed in this book. Then make sure the agency will *specifically customize a plan* that works best for *you*, employing all the tools *you* need — but none that you don't.

That's how reputable agencies work.

Using an internet marketing pro vs. doing it yourself

In many ways, we now live in a DIY society, and that's a good thing. With the expansion of 24/7 internet access into people's homes, we now have a wealth of information at our fingertips any time of day.

It's like having access to hundreds of thousands of libraries at all times, right there in your office, or living room, or porch, or wherever.

You can take your smartphone to the beach and access 100,000 times more information than a person could at the largest big-city library 25 years ago.

All that information makes it much easier for people to solve their own problems. Need tips for changing the pedals on your bicycle? There are not only step-by-step guides on the internet, but YouTube videos as well — and they're all completely free.

Some businesses take this approach for their internet marketing efforts. While a few are successful at this, that's generally because the business is related to internet marketing, or something closely affiliated with it, in the first place.

For most business owners, the best use of your time is doing what you do well. You handle hospice services, which is a very specific focus. You know how to work with individuals and families considering your services.

That's a lot on your plate. You should be focused on your clients and services, not trying to determine which keyword phrase gets you the best PPC results. You should be ensuring that a complicated burial service went exactly as planned, not figuring out how to edit image tags in your website code to improve your search engine rankings.

The DIY route might make sense if employing the right internet marketing firm were extremely expensive, but it's not. Some great firms are very affordable, and the increase in business will pay for the initiative many times over.

When it comes to justifying the cost, also consider this: Truly great marketing will help you become more selective.

The more potential clients you have, the choosier you can be. If you only have 10 clients (but can handle 25), you might have to undersell your services, because you can't risk losing the 10 clients you have. However, if you have 50 people who want to use your services, you can set your prices at a much more appropriate rate.

It's good to be selective. At Hospice Haven Marketing, quite frankly, *I'm* selective. Virtually every week, I turn down the opportunity to work a potential client or two.

If the numbers won't work for my business, it doesn't make sense to take on the campaign. (Remember, we have a 100 percent money-back guarantee, so we're taking on all the risk.)

I also turn down a potential client if that hospice's particular situation means it's unlikely to see a great benefit from our campaigns. I'm not in the business of taking people's money. I'm in the business of helping hospices grow and succeed. Anything less is unacceptable to me.

Red flags to watch out for when choosing an internet marketing firm

Now that you know the benefits of hiring a professional to handle your internet marketing, you need to know how to select one. The skills, experience, resources, and customer support abilities of various firms can vary widely from one to another.

Here I'll touch on additional "red flags" that are indicative of the way certain internet marketing consultants, particularly those with large firms, do business. I'll follow each one with a short explanation of how a legitimate business would handle the situation.

RED FLAG #1:
Conflicts of Interest

The problem with many large internet marketing firms is that while they are managing your campaign, they are also managing your direct competitors' campaigns.

You and your competitors all are competing for presence on the first page of search engine results. The space is limited. Whose interest does the firm have in mind?

In addition, at a big firm, your account representative is hired to sell — not to manage your campaign. How can he manage your campaign if he's pressured by his managers to hit sales goals every week? The rep's incentive is to sell, not to generate more sales for you.

If he triples your sales, he doesn't benefit it any way. If fact, if he ignores your account, he actually makes *more* money, because he's using the time he *should* be helping you to "close" more clients. Your account gets neglected and performance suffers.

At a legit marketing firm, your dedicated account manager is a dedicated marketing professional (at HHM, we call them *coaches*, because that's a big part of their role), not a salesperson. The managers are under no pressure to go out and find accounts. They're salaried and do not receive commissions. Their sole purpose is the continual optimization of your campaign, to bring in clients and make your phone ring.

They have top-level technical expertise in internet marketing and have helped numerous businesses — including hospices — generate more referrals and increase your average daily census.

RED FLAG #2
Lack of transparency

Most companies doing great work want to illustrate what they can achieve. That means sharing all the numbers with their customers. However, many large internet marketing firms do not provide you with all the data associated with your campaigns.

For example, they do not show you what exact keywords they bid on, which match type they use, what the bounce rate is, or the quality score associated with each keyword.

Those terms might be unfamiliar, but they're easily explained by a quality consultant. You don't need to know what all these words mean, but you do need to know that these are the words that measure whether a campaign is actually working.

A good internet marketer will not shy away from providing you a result with all this information. When you do good work for a client, you want to show it off. You want your clients to know exactly what you're achieving for them. When that information is hidden, you have no way to know.

It's equivalent to FedEx sharing its percentage rates for on-time arrivals. These are the key numbers that measure whether your marketing is succeeding or failing.

At a legit marketing firm, all the data is shared with you. The company should share with you exactly what it has done with your campaign every week. If a company does good work, why would it do anything else?

RED FLAG #3
Retention through contracts, not by results

A number of large internet marketing firms require contracts with minimum term commitments. Why is that a concern? It relates to how internet marketing works.

Business owners generally will know within 2–3 months whether their marketing provider is achieving the results they'd been promised. Unfortunately, if they want to terminate their campaign when they have a minimum term commitment, they're out of luck. Whether the campaign is soaring or nosediving, the business owner is on the hook.

If you're providing real, quantifiable results every month, why require a minimum term?

At a legit marketing firm, programs run month to month. Most partners see results in 30 days, but sometimes it takes 2–3 months, depending on the particular market.

After that time frame, you can terminate the service if you like. However, after 90 days, our clients typically are so thrilled with the results that they don't want to leave. They've achieved tremendous growth in their revenue and want to keep that going.

RED FLAG #4
Lack of technical knowledge

A lot of large internet marketing firms do not actually employ internet marketing experts.

They employ — as you'll remember from a few pages back — salespeople.

When you sign on with these firms, the person managing your pay-per-click campaign is not a marketing pro. It's your account representative.

Here's the problem: She's a salesperson, not a marketing technical engineer.

These account reps undergo a few weeks of crash training and are expected to manage your campaign. In fact, they are not certified.

Internet marketing is a very complicated process, and most salespeople are not very good at it. They earn their pay by being good at sales, not internet marketing.

At a legit marketing firm, your dedicated account manager is a dedicated marketing engineer. The managers' sole purpose is to *optimize* your campaign and get your phone ringing with calls from leads and referrers.

Unlike larger companies that check in on your account once a week, a legit firm will actively manage your account on a daily basis. It will monitor and adjust your campaign every day to achieve the outcome you and the expert decided upon together.

Once the firm has a great pay-per-click campaign set up for you, it can make sure it's always achieving results by working continuously to improve it.

Your target market's keywords might change every month. Your competition might suddenly be paying more for the same keywords, lowering your ad position. Perhaps your competition comes up with an ad that has a better offer — your ads still show, but fewer people click on them.

Continuous management is essential to keeping your online advertising costs down. That's what a good firm does.

RED FLAG #5
Focusing on the number of ad clicks, rather than on your business

With many large internet marketing firms, it's in their best interest to drive more clicks and spend all your budget each month, then come back to ask you to spend even more. The company makes its money based on the percentage of your spending.

If that sounds ridiculous to you, well, it does to me too. You're paying good money to drive potential clients and referrers to your website. It is critical that these visitors *go on to call your business*.

These companies do not do any website conversion optimization to ensure that traffic turns into referral sources and clientele. They only care about delivering traffic to your website, not broadening your referral network or increasing your average daily census.

Here's the problem with that: If your website isn't optimized in the first place — if it has a hard-to-find phone number, no offer, etc. — no amount of traffic will save you.

The primary purpose of the website is to create contact by phone or email. Optimizing your website for this purpose is a vital part of this process.

At a legit marketing firm, it all comes down to the bottom line: delivering a high return on investment. That means turning website visitors into sales.

Conversion rate optimization is an ongoing process of testing and tuning your website to make your phone ring. A good firm teaches you how to fine-tune your site to convert more online visitors into referrals and clients.

Conclusions

Staying ahead in a digital world

I've covered a lot of information in this book, and I realize it's a lot to take in.

Even without digging into all the minutiae of various marketing efforts in a digital world, as you can see, it's a very different battlefield than firms dealt with five years ago, or even two years ago.

I could tell you that everything will reach a nice, level plateau for a while, that once you understand how successful business marketing works in a digital world, that you'll be up on the status quo for the next several years.

But that isn't true.

The one constant of successful business marketing today is that nothing is constant. Social media networks that were incredibly influential at the start of this decade are virtually extinct today.

Mobile phones that were cutting-edge five years ago can't handle the vast majority of applications that are most used today.

Websites that attracted tons of traffic and converted visitors almost automatically are useless today (if they haven't been updated since then).

It all sounds scary, but in truth, the ever-changing influence of technology on local businesses in this digital age is a very *good* thing — for those willing to adapt and use it to their benefit.

These circumstances provide business owners who can adjust on the fly an enormous advantage over local competitors who are still doing business the 2020 way, much less the 2015 way.

I feel fortunate to have worked with lots of great local business owners just like you. I've seen firsthand how much more revenue they can generate by employing the very same tools I cover in this book.

These tools allow service providers to separate themselves completely from the pack. They can attract and target prospective referrers and clients long before their competitors are even aware of them. They can evaluate and measure their marketing approaches on the fly and improve them with virtually no delay.

Yes, things move fast these days. No, it will never be as simple as it was to simply run a local business. But that's simply the reality.

Times change, and people who accept that change and adapt to it will thrive. Those who do not, no matter how good they are at what they do, will stagnate and die off.

Perhaps that doesn't seem fair, and that's understandable. If you're the best in your location at providing hospice services, shouldn't that be all that matters? You do a great job. Admissions should naturally flow to you.

In truth, that's never been the case. Simply being good at your chosen vocation might be enough if you're an employee, but you're not an employee.

You run a hospice, which is a critical community service, but you're also a business owner. You took the plunge to be enterprising and run your own business, and that's wonderful. And as the owner, you know that whatever you do that running the business side is its own discipline, one that's very different from your day-to-day duties.

That discipline *did* change long before now.

It changed with the advent of websites and email. It changed when payroll went from being calculated with pencils and erasers to Excel spreadsheets. It changed when you went from contacting employees on pagers and walkie-talkies to smartphones.

Things are just changing faster now. Much, much faster. And it's not just the speed of things — it's also how far their influence has spread demographically. Incredibly powerful technology is in the palm of people's hands, and they're using it. Every age, every race, everywhere.

It doesn't matter if you're in the heart of a big city or working in a small town. Your potential clients are employing this technology in their day-to-day lives, and it's only going to become more pervasive.

This doesn't mean you need to become a master of the digital world. You can run your hospice and focus on what you do best. I've worked with lots of small business owners who just wanted the phone to ring.

And that's what I do. I employ the techniques in this book, in addition to other methods, as necessary, to attract visitors, and to convert those visitors into referral sources and clients.

The business owner can be as involved or removed as he or she likes. Some proprietors want to know and help control every aspect of the process. We can do that.

Some want to be almost entirely hands-off. We can do that too. And anything in between? Never a problem.

That's because you know best how to run your hospice. You know your strengths and weaknesses. You know what requires your direct supervision and what you can delegate. The same thing works with marketing in the digital world.

You can figure out what works best for you. A qualified marketing expert will adjust to your needs and expectations. The important thing is making your phone ring and your email inbox fill up with great leads.

Having read this book, you might think you need to employ all these marketing aspects immediately. On the other hand, it might sound so overwhelming, you might think it's best to ignore them altogether.

It should be obvious that the second option isn't an option at all. Other small businesses are employing these initiatives. They are getting more bang for their buck, expanding their client bases, and identifying the best and most profitable clients, using these methods.

With that said, your hospice doesn't necessarily need to do everything that's listed here. It depends on what you want to achieve. It depends on your budget. It depends on what resources are best directed to provide the very best return on your investment.

Various marketing companies do things different ways. My firm offers a 100 percent money-back guarantee, because we know what we do works. We know how to attract people who are looking for exactly what you provide.

Really, we're just like you: We're good at what we do, and we want to share our expertise with local businesses and help them grow.

Things move fast in the marketing world, and I don't expect the proprietors I partner with to stay up on these things. That's what *I* handle.

I stay apprised of every twist and turn and use that knowledge to benefit my partners. That's what a good marketing expert should do. It ensures that my clients stay ahead of the game, and it's a game that changes month to month, sometimes week to week.

Now that you've learned the basics of marketing in today's digital world, take some time to assess your marketing efforts and how they've been working for you.

Do you know how effective your advertising is?

Do you know what clients are saying about your hospice online, in user reviews, in ratings, and on social media?

Do you know whether people can easily access your website on smartphones?

Do you know whether your hospice even shows up on Google Maps or Apple Maps?

If you don't know the answer to these questions off the top of your head, it's probably time to take a long look at how you're marketing your hospice in the digital age.

Hospice owners who know the answers to these questions are highly visible to potential clients and referrers. Those who don't are typically flying way under the radar. You can't afford to be off that radar.

I want to thank you for taking the time to read this book and consider the points I've discussed here. If you're looking for more information on marketing your hospice in the digital age, I'll be more than happy to speak with you. Here are a few ways to get in touch:

Contact me anytime through email at:
welton@hospicehavenmarketing.com

Visit our website for hospice owners:
HospiceHavenMarketing.com

Call us toll-free:
(888) 383-2848

Thanks again, and let me wish you and your hospice great success now and for many years to come.

Welton Hong

Hospice Haven Marketing Information

Hospice Haven Marketing only makes money when our clients make money. Our work produces amazing results, so we stand behind it with a 60-day no-questions-asked money-back guarantee.

We have a simple evaluation process for businesses interested in partnering with us. We'll discuss your firm and how we can employ an online marketing campaign structured specifically for you.

Because we specialize in hospice and related fields, such as senior care and deathcare, we know exactly how to design a successful campaign for you.

If the numbers don't work out and we're unable to help you, it won't cost you a penny. That's why there's zero risk in contacting us for an evaluation. We want to make it easy for you.

Now is the time to act. You simply can't afford to get left behind. Learn how a professionally executed internet marketing campaign can make this your most successful year ever.

Visit our website for hospice owners:
HospiceHavenMarketing.com

Call us toll-free:
(888) 383-2848

Made in the USA
Coppell, TX
13 September 2026

89920725R00108